D1122971

CIVIL WAR BATTLES

THE MAPS OF JEDEDIAH HOTCHKISS

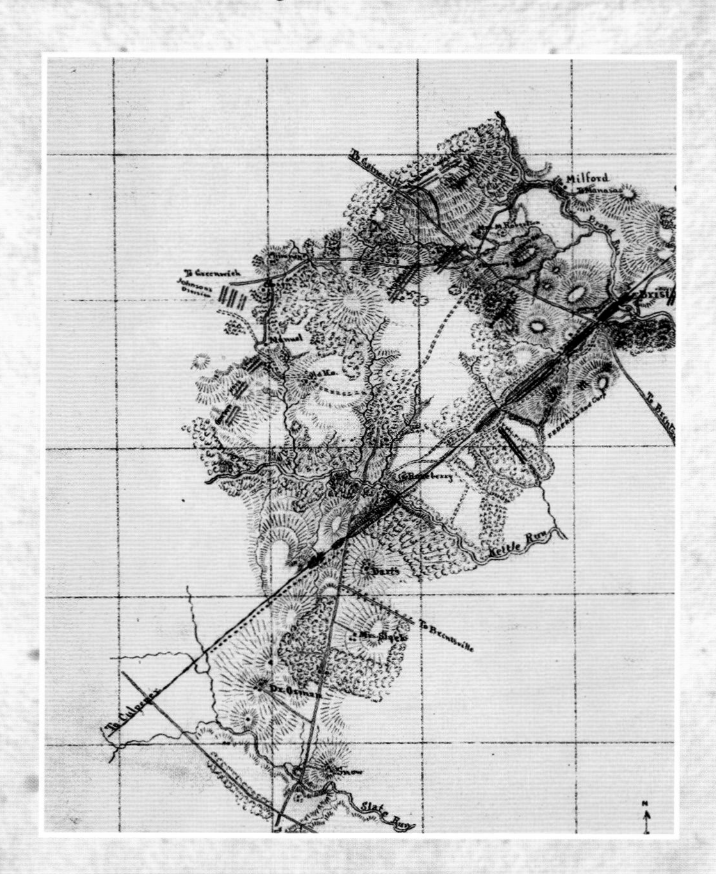

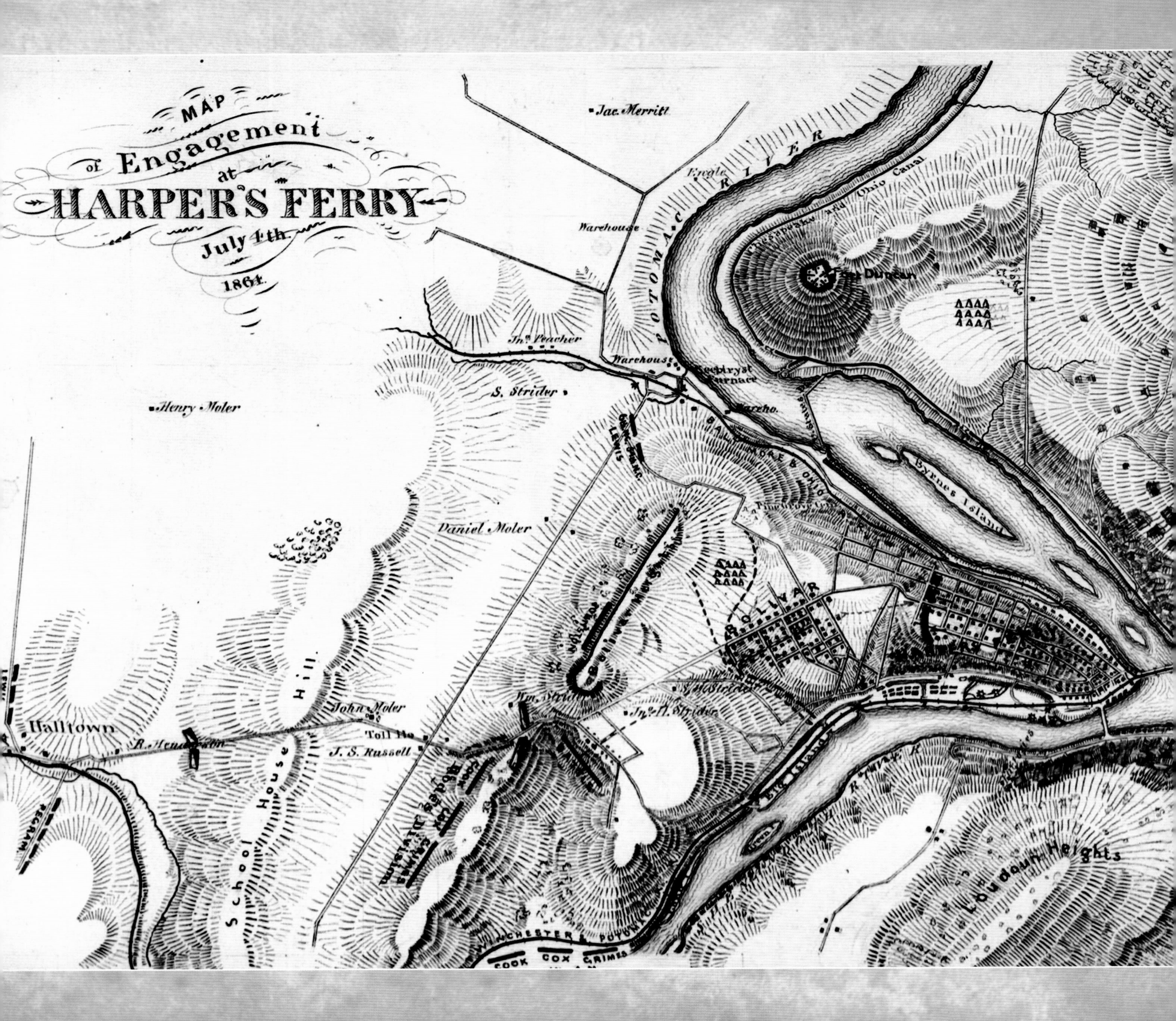
MAP
of Engagement
at
HARPER'S FERRY
July 4th.
1864.
Jac. Merritt
POTOMAC RIVER
Chesapeake and Ohio Canal
Eagle
Warehouse
Fort Duncan
Jno. Peacher
Warehouse
Furnace
S. Strider
Henry Moler
BALTIMORE & OHIO
Byrnes Island
Daniel Moler
BOLIVAR
School House Hill
John Moler
Halltown
Toll Ho
J. S. Russell
Wm. Strider
Jno. H. Strider
Rodes Division
Big Island
Loudoun Heights
WINCHESTER & POTOMAC
COOK
COX
GRIMES
PEGRAM

CIVIL WAR BATTLES

THE MAPS OF JEDEDIAH HOTCHKISS

Chester G. Hearn
Mike Marino

San Diego, California

THUNDER BAY
P·R·E·S·S

Thunder Bay Press
An imprint of the Advantage Publishers Group
10350 Barnes Canyon Road, San Diego, CA 92121
www.thunderbaybooks.com

All notations of errors or omissions should be addressed to Thunder Bay Press, Editorial Department, at the above address. All other correspondence (author inquiries, permissions) concerning the content of this book should be addressed to BlueRed Press Ltd., 6 South Street, Totnes, Devon, TQ9 5TZ, United Kingdom.

ISBN-13: 978-1-59223-952-8
ISBN-10: 1-59223-952-8

Printed in China
1 2 3 4 5 13 12 11 10 09

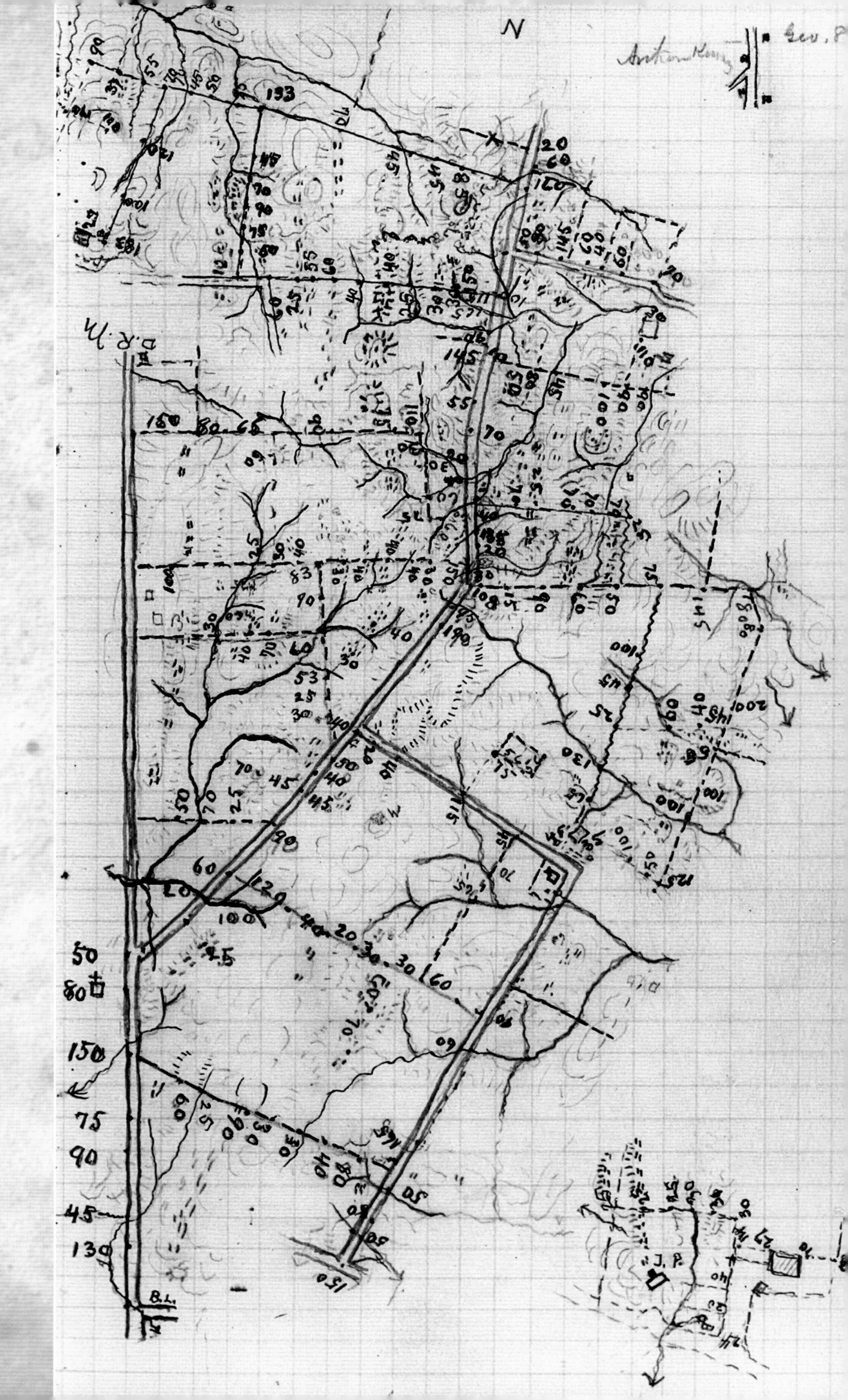

CONTENTS

INTRODUCTION

[Sketchbook Map #1-22, right half]
Hotchkiss settled in Virginia in 1847 after meeting Daniel Forrer, who lived in Mossy Creek and was searching for a bright young man to tutor his children. Hotchkiss took the job, and with plenty of time to follow his own interests, he began to study surveying, cartography, and engineering, which were mainly subjects taught in military schools.

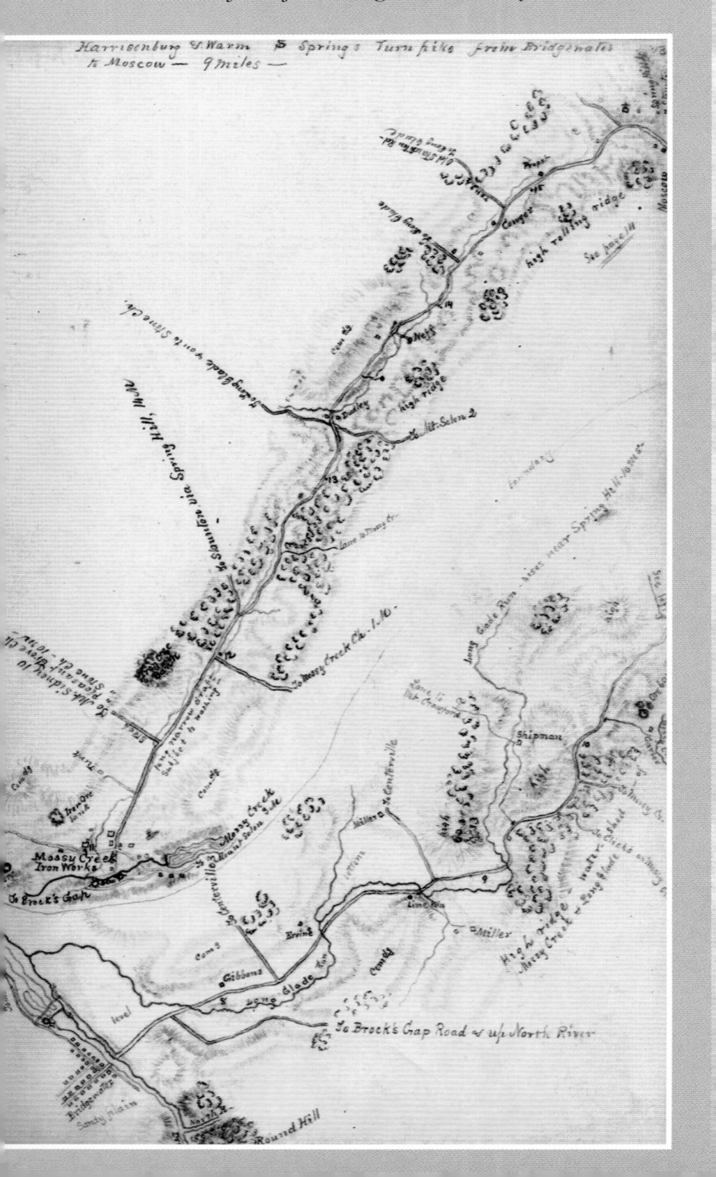

In 1861, when thirty-three-year-old Jed Hotchkiss rode into western Virginia to volunteer as a private, he stood about 5 feet 10 inches tall and weighed about 175 pounds. He had a fair complexion, dark hair with a few flecks of gray showing in his beard, and brown eyes that twinkled with a natural friendliness. A devout Presbyterian, Hotchkiss neither drank nor smoked. He loved life and he enjoyed laughter, but his joke telling seldom drew more than a polite guffaw. Perhaps this resulted from his talkative nature, which sometimes annoyed friends and military comrades, who came to him on business and for information and not for a long-winded social or theological conversation.

Though not a native Virginian, Hotchkiss moved to the Shenandoah Valley from Windsor, New York, in 1847 and earned his living as a schoolteacher, which might explain his loquaciousness. In 1853 he married Sara Ann Comfort, a Pennsylvanian who gave him two daughters, Ellen May (Nelly) and Anne Lydia. Nelson Hotchkiss, Jed's brother, also settled in the valley, and in 1859 the two men founded Loch Willow Academy, a school for boys near Churchville. The Hotchkiss family quickly fell in love with the beautiful Shenandoah Valley, and were devoted and loyal citizens of the Commonwealth of Virginia for the rest of their lives.

Hotchkiss studied botany and geology at Windsor and planned to be an educator. He aspired to living at peace with God, teaching school, owning a few acres of land for grazing horses and cows, and planting an ample garden to help feed his family. Fighting a war against New Yorkers and Pennsylvanians did not fit into his plans for living quietly in a pastoral setting. Yet Hotchkiss did fight, but not as a musket-bearing infantryman or a saber-

Area of Operations of Jed Hotchkiss 1861–1865 [Map #1]
For Jed Hotchkiss, the Civil War began during the summer of 1861 at Camp Garnett in western Virginia. From there it spread to the Shenandoah Valley, Richmond, Manassas, Sharpsburg, and after looping back into Virginia, reached as far north as Harrisburg, Pennsylvania. In 1864 the final campaign brought Hotchkiss to the outskirts of Washington, D.C.

The date of this photograph is unknown, but it was probably taken during the war when Hotchkiss was in his mid-thirties.

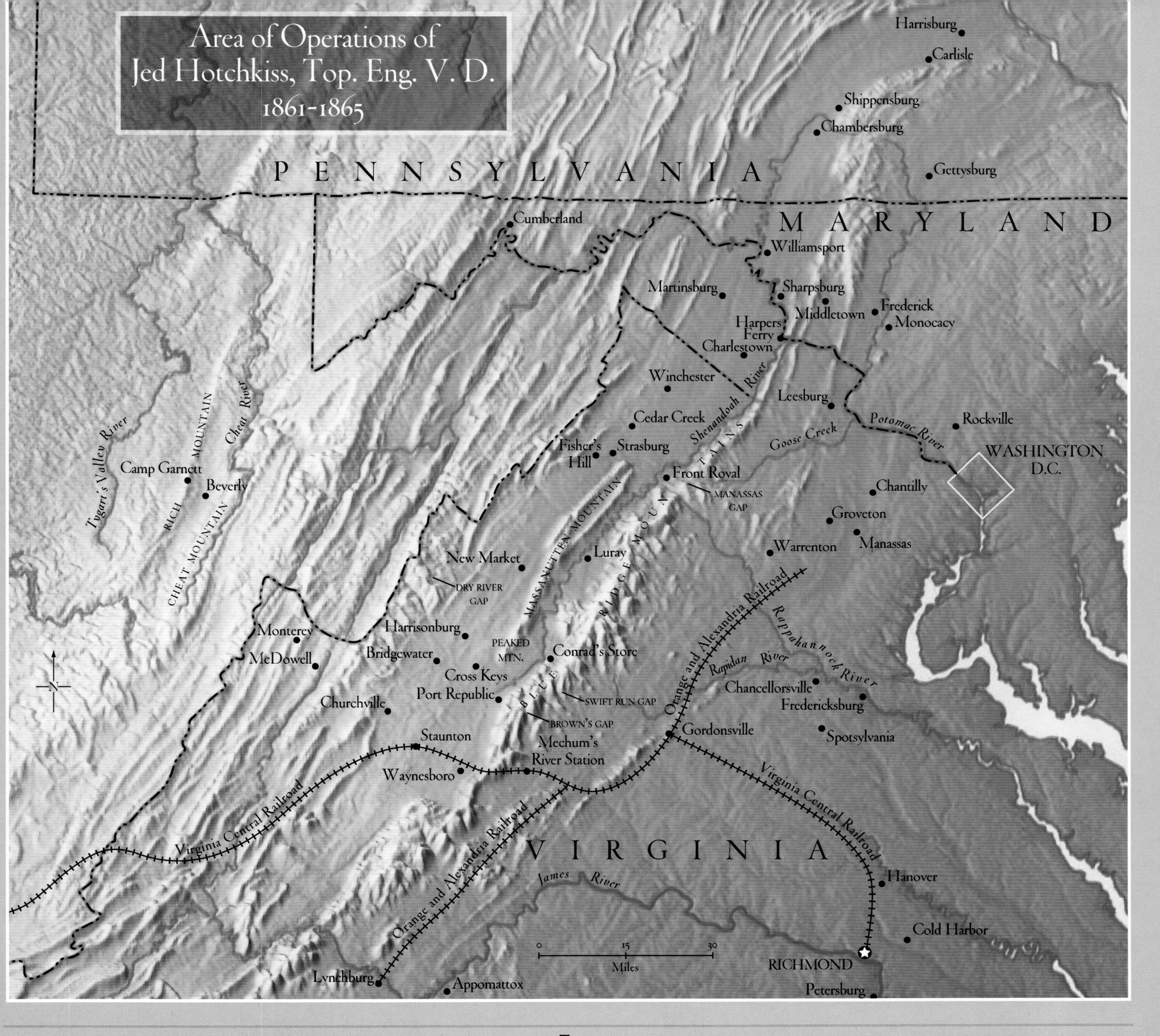

Area of Operations of
Jed Hotchkiss, Top. Eng. V. D.
1861-1865
PENNSYLVANIA
MARYLAND
VIRGINIA
Harrisburg
Carlisle
Shippensburg
Chambersburg
Gettysburg
Cumberland
Williamsport
Martinsburg
Sharpsburg
Middletown
Frederick
Monocacy
Harpers Ferry
Charlestown
Winchester
Leesburg
Cedar Creek
Fisher's Hill
Strasburg
Front Royal
MANASSAS GAP
Rockville
WASHINGTON D.C.
Chantilly
Groveton
Manassas
Warrenton
Potomac River
Goose Creek
Shenandoah River
Cheat River
Tygart's Valley River
RICH MOUNTAIN
CHEAT MOUNTAIN
Camp Garnett
Beverly
New Market
DRY RIVER GAP
Luray
MASSANUTTEN MOUNTAIN
BLUE RIDGE MOUNTAINS
Monterey
McDowell
Harrisonburg
PEAKED MTN.
Bridgewater
Cross Keys
Conrad's Store
Port Republic
SWIFT RUN GAP
BROWN'S GAP
Churchville
Staunton
Mechum's River Station
Waynesboro
Gordonsville
Chancellorsville
Fredericksburg
Spotsylvania
Rapidan River
Rappahannock River
Orange and Alexandria Railroad
Virginia Central Railroad
James River
Hanover
Cold Harbor
RICHMOND
Petersburg
Lynchburg
Appomattox
0 15 30
Miles
N

rattling cavalryman. He fought with a sketchbook, pencils, pen and ink, a pair of binoculars, and simple surveying instruments.

To make a map, Hotchkiss first studied the physical features of an area, sketching the elevations and depressions and noting streams, bridges, roads, hills, houses, mills, cow pastures, grain fields, and woodlands. Sometimes he used a rope or chain to record distances, and sometimes he walked the ground in measured steps, taking into account undulations in the landscape and marking or staking off small increments every hundred paces as he progressed. In normal daylight, he could accurately estimate distances up to a mile by eye and come within a few hundred feet.

Hotchkiss's surveying kit contained an aneroid barometer, which was a round, moderately expensive instrument about ten inches in diameter that fitted into a wooden box. The barometer had a dial that recorded atmospheric pressure in relation to sea level. Atmospheric pressure decreases in relation to sea level as elevation increases, and by moving higher or lower, Hotchkiss could measure the heights of hills in relation to ground level by setting his position in relation to sea level and making calculations from the aneroid barometer's dial recordings. Hotchkiss also used a lensatic pocket compass with collapsible sights that helped him find bearings and set courses. Walking a stretch of road rarely resulted in a straight line, and the compass allowed the surveyor to turn onto different roads and make corrections in distances. Hotchkiss also carried a field book for making sketches and notations, and some of those sketches eventually became parts of maps. He later captured a long Federal coat and hired a seamstress to add deep pockets for everything he carried.

Hotchkiss drew every map at least twice, first in pencil so corrections could be made, and then in ink. Over time, as he collected more information, the map went through a series of modifications, which eventually led to a more comprehensive map. Many of the maps made by Hotchkiss were works in progress, sometimes for many years. His self-taught topographical expertise and the maps he produced changed his life forever, and the experiences written in his journal and correspondence form an important record of the Civil War.

MOSSY CREEK

Impressed with Hotchkiss's teaching ability, the Forrer family raised money from other local residents and in 1852 built the Mossy Creek Academy after Hotchkiss agreed to serve as principal. The setting could not have been more delightful for Hotchkiss. He lived near Mossy Creek Lake, established a field laboratory for advancing his botanical, geological, and engineering interests, and in 1853 joined the Mossy Creek Presbyterian Church.

Hotchkiss did not date his sketch of the Mossy Creek area, where he lived until 1858, but the map appeared in the first of his eight sketchbooks. It could have been made when he lived at Mossy Creek, or it could have been made during the war. The sketch shows the route to Staunton, where he later spent most of his life, and all the side roads and farm roads that became so important during the battles fought in the Shenandoah Valley. Streams, houses, hills, gaps through the mountains, woods, fields, towns, and where roads led all became a typical part of Hotchkiss's wartime maps. Although the sketch has many similarities to maps made by Hotchkiss during the Civil War, it is also possible that this sketch was made while he lived in Mossy Creek.

In 1860 most Americans lived east of the Mississippi River and had little need for detailed maps. To travel long distances, people depended on trains, steamboats, and coaches. Roads were poorly marked, and most commercially produced maps were out of date and not always obtainable. Although the government engaged in mapping for official use, nobody anticipated a need for military maps. When Southerners decided to experiment with secession, they did so without accurate maps.

In 1861 Hotchkiss received a lieutenant's pay without serving as an officer in the Confederate army. In 1862 he received a captain's pay, once more without officially becoming an officer. In 1865 he received a major's pay and still never became an officer, although his fellow officers called him "Major." He remained a civilian throughout the war and never wore a uniform, but everyone in General Robert E. Lee's Army of Northern Virginia knew who he was. Southern soldiers observed him in camp, always with a sketchbook in his hand and a packet of writing instruments in his long black coat or tucked in his pocket. They would see him perched on a hill, perhaps sitting on a rock and deeply engrossed in some far-off object. In camp they would find him in his tent, bent over a table with officers coming and going, and sometimes at dusk they would see him riding into camp with one arm cradling his sketchbook and the other navigating his horse. Sometimes he traveled country roads in his wagon, wearing a slouch hat and looking like a local farmer going about his business.

During General Thomas J. "Stonewall" Jackson's brilliant Shenandoah campaign—which this narrative covers in depth—word began to spread through the camp that Hotchkiss was not merely a wandering sketcher of terrain but an intelligent and proficient engineer upon whom the general depended for developing his campaign strategies. Although Jackson had earned the nickname "Stonewall" at First Manassas in July 1861, the allusion came from Brigadier General Barnard E. Bee, a fellow brigade

Shenandoah Valley [Map #147]

While Hotchkiss's Shenandoah Valley map is too large for book-size publication, he and his staff made dozens of partial reductions for division and brigade commanders. Some of the reductions became base maps covering a strategic area, to which other reductions and enhancements occurred for battle reports and campaign planning. This particular segment covers the top third of the Shenandoah map. The reduction clearly shows the network of roads, waterways, towns, and topographical features from Winchester to Fishers Gap. Every box on the original map was one inch square and represented approximately 1.25 square miles.

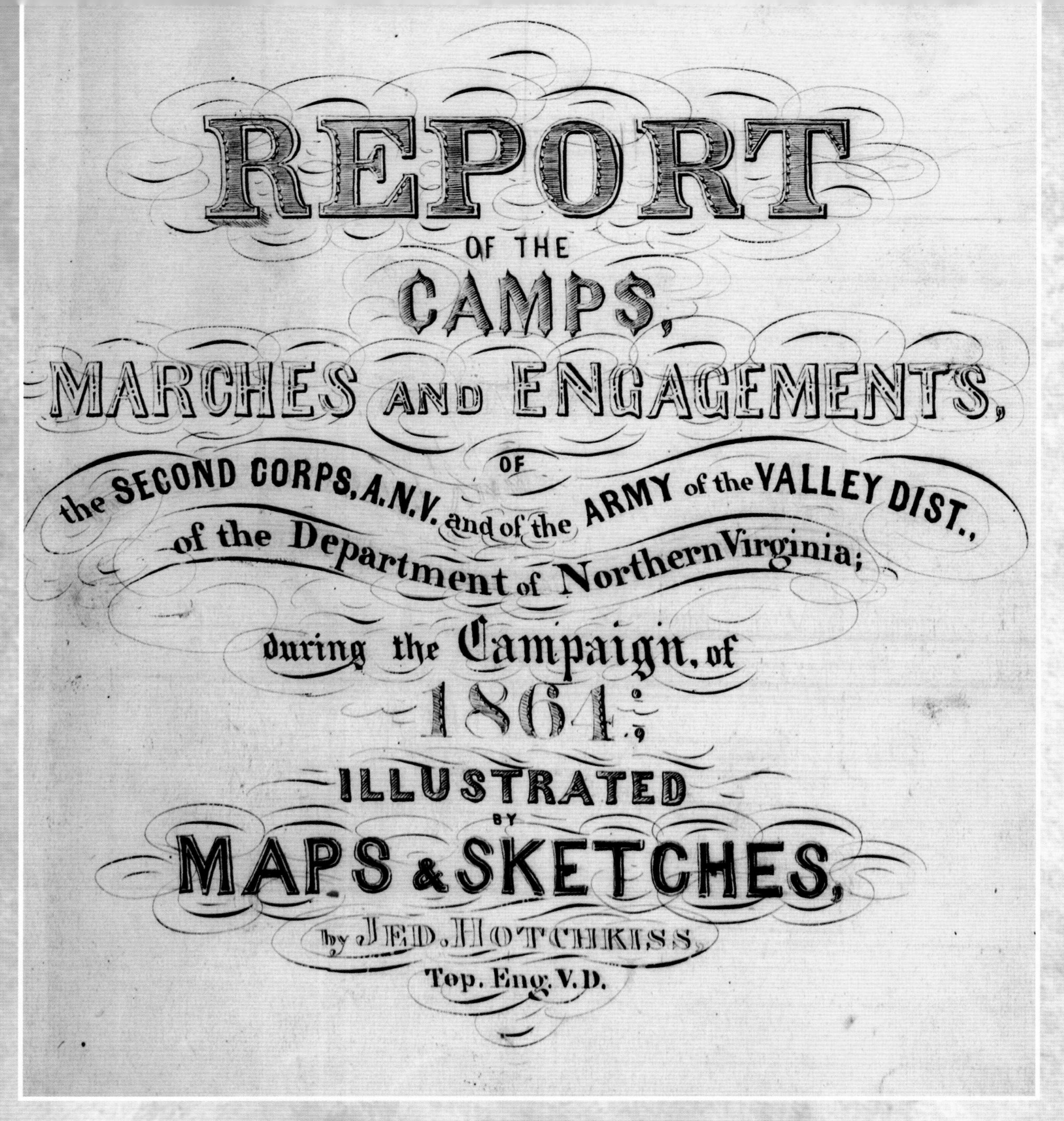

Cover for Sketchbook #8

Sketchbook #8 contains page after page of refined sketches produced by Hotchkiss and his staff during the campaigns of 1864. Most of the sketches in the first seven books consisted of small pieces of topography or actions that later went into one of his maps. Only Hotchkiss knew the purpose of most of his sketches because few of them identified the location where they were drawn. He sometimes put six to eight small sketches on a single page, and other pages of his sketchbooks carried descriptive notes only he, or possibly his staff, could interpret. During the Civil War, seven people assisted Hotchkiss at different times, but Samuel B. Robinson and C. William Oltmanns became permanent members of his staff in 1864.

commander, who with Jackson defended Henry House Hill. Bee died of wounds the following day, denying history from knowing whether he meant that Jackson was standing like a stone wall and doing nothing or whether Jackson was standing like a stone wall and valiantly holding his ground. History gave Jackson the benefit of the doubt, but the general's next campaign fared poorly in western Virginia. Jackson didn't begin to distinguish himself as a great military tactician until Hotchkiss joined his command and began to identify routes that led to victories in the Shenandoah Valley, at Second Manassas, during the Antietam campaign, and eventually at the battle of Chancellorsville. As venerated Civil War historian Douglas Southall Freeman declared in *Lee's Lieutenants*, "With Hotchkiss away, Jackson was not blinded, but his vision was dimmed." After Jackson's death, generals Robert E. Lee, Richard S. Ewell, and Jubal A. Early all relied on Hotchkiss for accurate maps and for plotting routes to beguile the enemy. At least some of the credit for the Union's frustration in defeating General Lee's bedraggled Army of Northern Virginia belongs to Hotchkiss.

As a self-trained surveyor with an interest in geology and botany, Hotchkiss developed an unrivaled eye for terrain, which he demonstrated in 1861 while serving briefly in western Virginia. When he joined Jackson's Valley Army the following March, the general tasked Hotchkiss by asking him to produce a map of the Shenandoah Valley. The resulting masterpiece of cartography, drawn to a scale of 1:80,000 on tracing linen, measured seven and a half by three feet and set the stage for three major campaigns in the Shenandoah Valley from 1862 to 1864. Pasted together in sections that fit precisely, the map remained with Hotchkiss throughout the war, and he laid it before commanders before and during every important campaign in the Shenandoah Valley.

SHENANDOAH VALLEY

Once Hotchkiss mapped an area, his memory functioned like a modern-day computer storing files and bits of photographic information for spontaneous retrieval. When preparing for a battle, he would return to camp after performing a reconnaissance and, using his sketchbook and a pencil, show a conference of generals the encampments, positions, and movements of the enemy; the routes to use in deflecting or performing an assault; and answer their questions with knowledge, sketches, and accuracy. He sometimes sketched upside down so that officers sitting across from him could read a map as he explained it. During a battle, he often roamed the field noting the position of opposing forces so that at a later time he could produce an after-action map for official reports going to the Confederate war department. There were also battles he missed, like First Manassas (First Bull Run). A year later, he spent time on the field during Second Manassas and began a credible map of the first battle.

During the Civil War, Hotchkiss produced at least eight sketchbooks. Freeman borrowed from the sketchbooks when writing an entirely new account of Jackson's Shenandoah Valley campaign in *Lee's Lieutenants*. All the early authors of the Army of Northern Virginia's history tapped the Hotchkiss collection for maps to use as reference. His collection contains some 600 maps and sketches relating to Virginia, West Virginia, Maryland, and Pennsylvania during the Civil War. More maps were added after 1867 when Hotchkiss opened an office in Staunton, Virginia, as a topographical and mining engineer. The office remained in business until Hotchkiss's death in 1899. It is a selection of his wartime maps and sketches that provide the foundation for this book.

The maps provide a chronology of the war witnessed by Jed Hotchkiss and of the men and officers who fought from the early battles in western Virginia and in the major battles of the Army of Northern Virginia to the very end of hostilities. Many of the maps are too large to fit the pages of the book, but they have retained their legibility, so a small selection has been inserted as loose maps. More than forty-five additional maps and sketches are included here, and they clearly demonstrate how Hotchkiss developed from a rough sketcher in the first chapter to a refined and artistic mapmaker in the latter chapters.

The book is about more than maps. The pages chronicle the important battles of the Army of Northern Virginia in which Hotchkiss participated. The actual fighting that took place becomes the central feature of every map. The book is a blend of the actual battles and campaigns, and of the self-trained topographer whose maps, by most accounts, extended the war. Jackson's Shenandoah Valley campaign remains one of the most remarkable military operations in modern warfare, as does his movements at Second Manassas in August 1862, the capture of Harpers Ferry in September prior to the battle of Antietam, and his unparalleled flank attack on the Army of the Potomac during the battle of Chancellorsville in May 1863. Jubal Early's desperate Washington raid during the summer of 1864 also stands as an outstanding feature of this book.

The Civil War changed the lives of every American, including Jed Hotchkiss. More than 600,000 soldiers died in the conflict, and those who lived to see the country reunified emerged from the war to resume peaceful careers. Hotchkiss came home a different man and rebuilt his life from what he learned in the war. Others did the same, but Jed Hotchkiss earned a special place among the millions who served.

For a man who carried a sketchbook into battle and never a musket, Jed Hotchkiss left an indelible mark on the operations of the Army of Northern Virginia. Douglas Freeman, one of the most distinguished authorities on the Civil War, correctly observed that Hotchkiss's contribution to the Southern cause merited "a place in Confederate service almost unique."

CHAPTER ONE: THE OPENING BATTLES

On April 17, 1861, the arrival of a telegram at Churchville, Virginia, hailed the commonwealth's secession from the Union. A teacher from Hotchkiss's Loch Willow Academy rapidly organized an infantry company and recruited several pupils from the school. Other students enrolled in a cavalry company. Facing the inevitable, and deprived of his principal means of livelihood, Jed Hotchkiss closed the academy and offered his services to the Confederate army.

On July 2, 1861, Jed Hotchkiss rode into western Virginia with a small bundle of surveying equipment and stopped at Beverly, approximately fifty miles west of the Shenandoah Valley. He reported to Lieutenant Colonel Jonathan M. Heck, who was at a loss about what to do with a schoolteacher equipped with surveying instruments and temporarily assigned him to a group of teamsters. A day later, Brigadier General Robert S. Garnett arrived to organize 4,500 Confederates for defensive operations in the neighborhood of Rich Mountain, which was a broad area occupied by 15,000 Federals under the command of Brigadier General George B. McClellan. Having no organizational affiliation, Hotchkiss joined the militia company of Captain Robert D. Lilley's Augusta-Lee Rifles. Before Hotchkiss began drilling with the company, Garnett detached him on July 3 with instructions to survey the region. Hotchkiss had nearly come to the end of his first military topographical assignment when on July 11 four Federal regiments under the command of Brigadier General William S. Rosecrans struck Colonel John Pegram's Twentieth Virginia Infantry Regiment astride the Buckhannon–Beverly road.

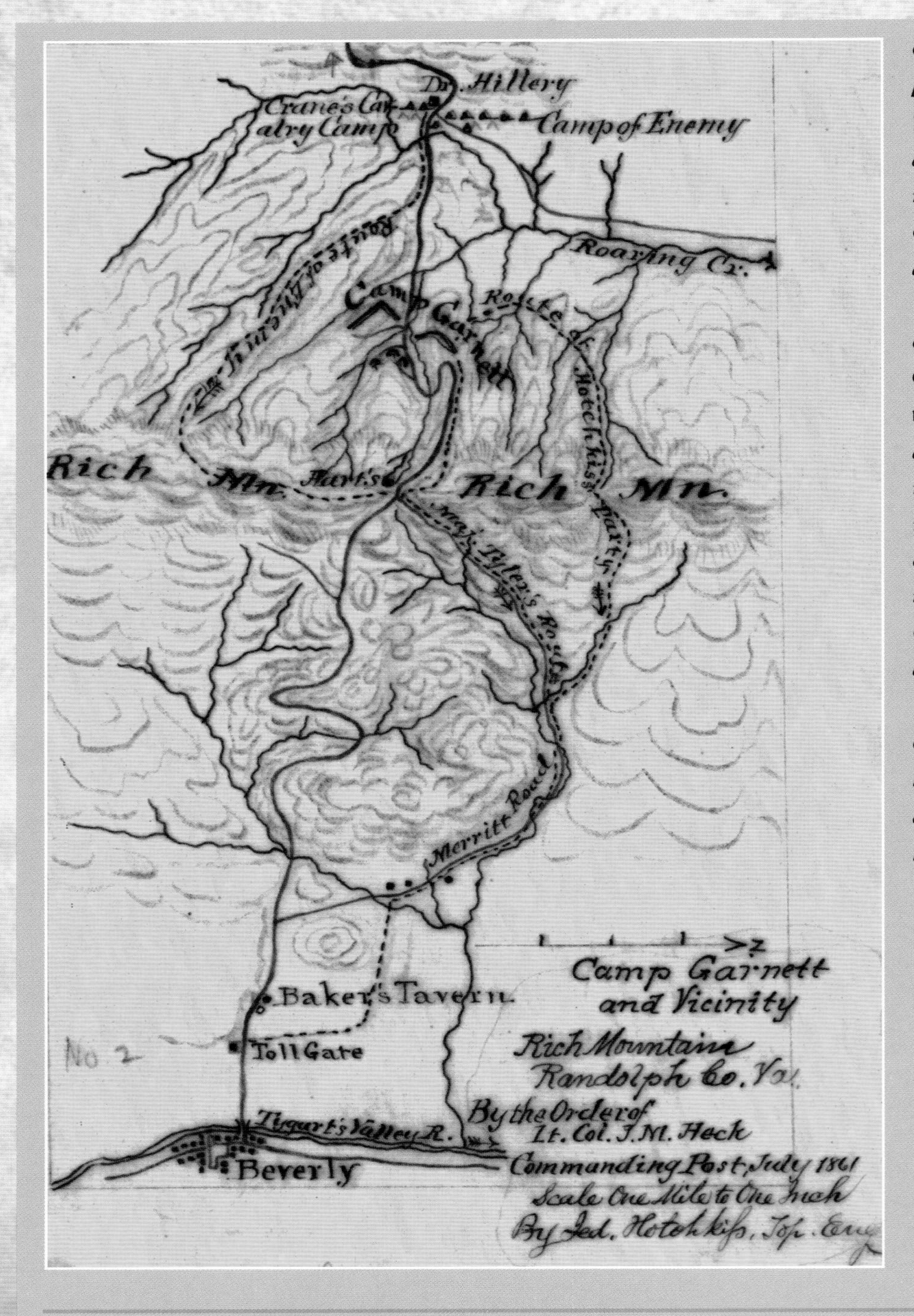

Camp Garnett
[Map #70]
The map of Camp Garnett and vicinity, made at the request of Lieutenant Colonel Jonathan M. Heck, appears to be the first Civil War map in the Hotchkiss collection. From The Official Military Atlas of the Civil War, *it also appears that Colonel Heck retained two of Hotchkiss's maps in his personal collection: a refined version of the Camp Garnett vicinity and another of the western Virginia area of operations. The actual location of Camp Garnett appears near the top of the map.*

THE BATTLE OF RICH MOUNTAIN: JULY 11, 1861

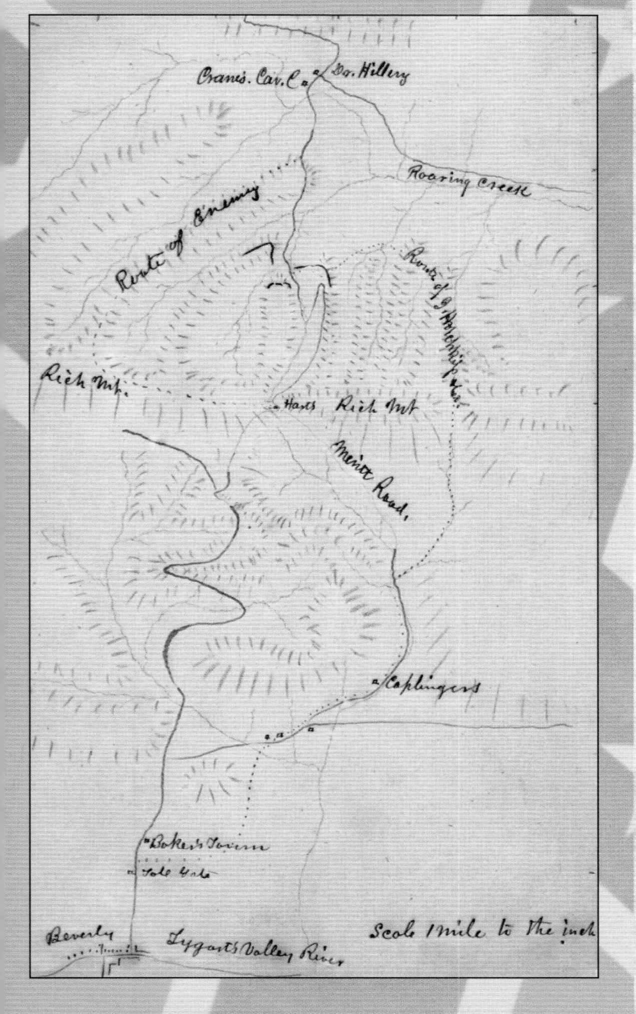

When comparing the sketch of Rich Mountain to the above map of Camp Garnett and vicinity, many features are similar. This was typical of Hotchkiss's work. He built his maps from crude sketches, and later produced a highly readable map. Most of the fighting occurred around Hart's House, which is located in a gap near the center of Rich Mountain through which the Buckhannon road passes.

Hotchkiss finished the sketch while sitting in a trench, fully aware that Camp Garnett lay on one end of Rich Mountain while the main Confederate force lay some fourteen miles away at the opposite end of the mountain. Around midnight, Pegram ordered Heck to assume command, withdraw the men along the length of the mountain, and rejoin Garnett. Heck, who had not studied the area, asked Hotchkiss to lead the way. The schoolteacher hurriedly formed the Augusta-Lee Rifles into a column and began the midnight trek up and along the ridges of Rich Mountain. It soon became apparent from noise below that a Federal column was also moving along the slope but in the opposite direction. Having taken a prisoner, Hotchkiss learned the Federal countersign, which was "Indian." When he heard a soft whistle ahead, he replied with the countersign. The column moved swiftly along a logging trail, avoided the enemy, and at dawn reached safety. In the first rays of daylight, Hotchkiss found that only the Augusta-Lee Rifles had followed him. Pegram had recalled the rear of the column when he became aware of increased Federal activity around midnight.

On that day, Hotchkiss became Colonel Heck's unofficial mapmaker. Not all the maps made while working for the colonel went into Hotchkiss's personal collection. After the war, some began to surface from Heck's documents when authorities started to assemble the military maps.

A section of the turnpike to Philippi stretched from Beverly to Leadsville and lay in the Tygart River Valley. The valley separated two long and slender ridges, Rich Mountain on the west and Cheat Mountain six miles to the east. General McClellan deployed a 12,000-man Federal division in order to attack Garnett's force of 4,500 widely scattered grayclads. McClellan made the mistake of believing that his much superior force faced more than 10,000 Confederates. His chronic habit of overestimating the strength of the enemy would bedevil him throughout the length of the war.

On July 9, 1861, McClellan ordered Brigadier General Thomas A. Morris to concentrate his brigade in front of a ridge situated between the northern end of Rich Mountain and Laurel Mountain and to attack Garnett's position on a signal from General Rosecrans. McClellan had previously detached Rosecrans with 2,000 men and sent him to the opposite end of Rich Mountain, which was about twelve miles away.

In a heavy rainstorm, Rosecrans moved silently into position. By some inexplicable blunder, a bugler called reveille and assembly at midnight. The noise alerted Pegram, and he made hurried adjustments to counter an attack. Dividing his 1,300 troops, Pegram sent 350 men and a single cannon to guard the road to Rice Mountain's summit. With his main force of 950 troops and three cannons, Pegram put his grayclads to work on cutting trees and erecting a breastwork of logs. Rosecrans decided against a frontal attack and made a long, circuitous movement. After ten hours of strenuous marching, he struck Pegram's original campsite and found it deserted.

Rosecrans's forces eventually came into the open early in the afternoon and received a dose of Confederate artillery and musket fire. After more than two hours of skirmishing, a section of Federal infantry charged and broke through the Confederate line. Pegram tried to reinforce his position by recalling the troops placed on Rich Mountain, but they turned away and scrambled back up the slope.

Rosecrans dallied because he had heard nothing from McClellan, who was supposed to be assaulting Garnett on the other end of the mountain. McClellan had heard cannon fire throughout the afternoon and worried that Rosecrans had been defeated, so he did nothing. A few of Pegram's men escaped over the mountain and began to appear in Garnett's camp,

Brigadier General William Wing Loring (1818–86) *gave Jed Hotchkiss his first opportunity to work for a general. The novice mapmaker spoke highly of Loring, but several months later the general had an altercation with Stonewall Jackson over disobedience of orders and left the department.*

warning that Rosecrans was not far behind. Pegram, however, remained with his command. Finding his force surrounded by Federals, he surrendered with 554 of his men.

Garnett evacuated his camp and, on reaching the turnpike, attempted to withdraw from the area by turning off the road at Leadsville and vanishing through a gap below Cheat Mountain. Rain slowed Garnett's movement. After he interposed a delaying action at Carrick's Ford with a rear guard of ten Confederates, McClellan's troops caught up with Garnett and killed him at another ford a short distance away.

Hotchkiss was able to escape, but lost most of his engineering equipment. He filed a claim with the Confederate government for one aneroid barometer, one set of mathematical instruments, and two sets of compasses, altogether totaling $83.

THE GENERAL'S CARTOGRAPHER

The death of Garnett produced a sobering effect on the Confederacy. After the battered brigade moved to Monterey, forty-two-year-old William W. Loring arrived to assume command. Loring had lost an arm in the Mexican War and his watery eyes bulged from under an oversized forehead. The combination, according to Hotchkiss, gave the general the appearance of a worn-out officer gripped in sadness.

Loring soon discovered he had no maps. On learning that Hotchkiss had sketched Rich Mountain and led an escape over the ridge, Loring put him to work on mapping the entire area. Hotchkiss's early opinion of Loring instantly improved. Named acting adjutant, he now worked for a general, which had been his goal.

Sanitation problems at Camp Monterey distressed Hotchkiss, who wrote, "It is the worst place for men to be who are not well . . . some of them die daily." He also complained of personal illness but continued to perform his duties. The Federals had not threatened since the battle at Rich Mountain, and the temporary pause in hostilities resulted from General McClellan being called to Washington to organize the Army of the Potomac, leaving Rosecrans in command.

Much to Loring's chagrin, President Jefferson Davis detached General Robert E. Lee from his duties in Richmond and sent him to western Virginia to recover the territory lost by Garnett. The general's arrival offended Loring, who stubbornly resisted Lee's efforts to attack Rosecrans. Both generals finally agreed they should have a map of the Tygart Valley before commencing offensive operations. Appointed an unofficial lieutenant of engineers for payment purposes, Hotchkiss returned to the Tygart Valley in early August. He spent countless hours under a soaked umbrella sketching a map that highlighted all the topographical features of the landscape, including roads, streams, ridges, farms, towns, and enemy activity for Lee's campaign. He returned to camp with typhoidlike symptoms and laid the map before the generals. When the army moved out on September 8, Hotchkiss followed, although he was too sick to be of much use.

Brigadier General Richard Brooke Garnett (1817–63) *fought well during the battle of Kernstown but withdrew his brigade without orders and was unjustly relieved by Stonewall Jackson.*

Western Virginia Theater [Map #66]

While assigned to Lieutenant Colonel Jonathan Heck, Hotchkiss began to sketch the western Virginia theater of operations in which he found himself involved. He lived near Churchville, a few miles northwest of Staunton (lower right) and on the road to Monterey, which he would have taken to reach Cheat Mountain and, beyond it, Rich Mountain. An interesting feature of Hotchkiss's maps is the number of times he wrote something upside down, which gave the person sitting across from him the ability to read the map while Hotchkiss read and explained his notes from the opposite side.

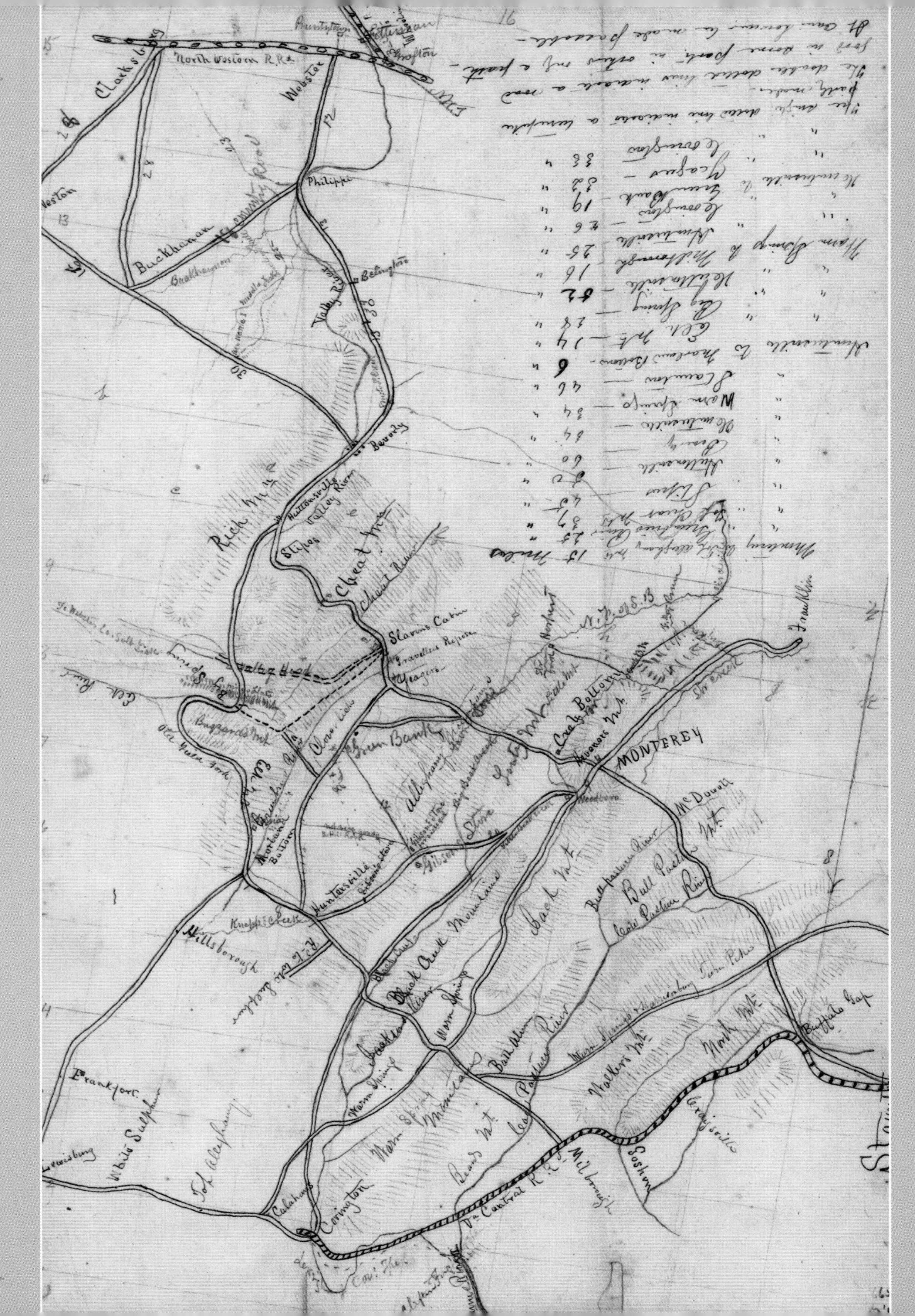

THE BATTLE OF CHEAT MOUNTAIN: SEPTEMBER 11–15, 1861

"Oh! Brother, I never want to spend such a day again."

—FROM HOTCHKISS'S PAPERS, SEPTEMBER 18, 1861

Cheat Mountain, though a minor battle, became General Lee's first military operation of the Civil War. The Confederates had been inactive since mid-July, and Lee expected General Loring to advance and drive the Federals off Cheat Mountain. The area held strategic importance in western Virginia because it contained mountain passes with roads linked to the Staunton–Parkersburg turnpike, which stretched from the Shenandoah Valley to the Ohio River. When Hotchkiss reconnoitered the area, he noted that Rosecrans had split Brigadier General Joseph J. Reynolds's command, placing one regiment under Colonel Nathan Kimball on Cheat Summit and five regiments under Reynolds at Elkwater in the Tygart Valley. The two Union wings were vulnerable, separated by seven miles by following mountain trails and eighteen miles using roadways.

Lee took personal command of the 15,000-man Confederate division and split it into five brigades. Having learned from Hotchkiss of a concealed route to Kimball's position, he ordered Colonel Albert Rust to strike Kimball on Cheat Summit. He ordered Brigadier General Samuel R. Anderson's brigade to occupy the wagon road to the summit while Brigadier General Henry R. Jackson's brigade occupied the remainder of the summit. Against the larger Federal force more than seven miles away, Lee intended to attack with his three remaining brigades, two of which Loring directly commanded and the other which he nominally commanded.

The complicated and tightly coordinated plan depended on Rust occupying the summit on schedule, after which assaults would commence throughout the valley.

The march began on September 10 in dreadfully wet weather conditions. The following day, Loring encountered stubborn resistance from Reynolds, who withdrew to Elkwater. On September 12 Loring engaged Reynolds at Elkwater and Anderson cut the wagon road to the summit. The crucial surprise attack on Cheat Summit, on which everything depended, never materialized. Rust stopped when a handful of captured Federals convinced him that Kimball had 4,000 men entrenched on the summit when there were, in fact, only 300.

With Rust still paralyzed on September 13, the Confederate attack began to wilt in the rain. Reynolds reinforced Kimball with three Union regiments, thus blocking Rust. Because Hotchkiss had fallen too ill to participate in scouting, Lee sent a detachment under Colonel John A. Washington to reconnoiter Elkwater. A Federal bullet took Washington's life during a skirmish near the Union line. With the element of surprise gone and the Federals in control of Cheat Summit, Lee withdrew on September 15 during a heavy rainstorm and ingloriously returned to Monterey. Dismayed by Lee's failure, President Davis recalled the general and sent him to South Carolina to oversee fortifications.

WAITING FOR SPRING

Hotchkiss stumbled into Monterey feverish, exhausted, and depressed over the Cheat Mountain fiasco. His head throbbed, his joints ached, the world seemed blurred, and he could not sleep. He watched glumly as men with the same symptoms collapsed and died. No longer fit for service, he longed to be home where his loving wife, Sara, could nurse him back to health. General Lee observed Hotchkiss's condition and sent him home. Granted leave after a seventy-nine-day campaign that ended in defeat, Hotchkiss reached home completely used up and went to bed. During October, as his health gradually improved, a letter arrived from the government confirming that he would receive a monthly pay of $93.33 for services rendered, the civilian equivalent of a lieutenant of engineers. The notification boosted his spirits. He had witnessed war, made friends, and contributed his talents. News of the Confederate victory at Manassas (First Bull Run) in July 1861 made him anxious to return to active service. Still in poor health, he decided to wait for spring.

During the unusually harsh winter of 1861–62, the inhabitants of the Shenandoah Valley learned of Major General McClellan's massive buildup of troops in camps outside Washington. In Hotchkiss's opinion, McClellan's vaunted reputation emanated from skirmishes with a small force of Confederates at Rich Mountain during the past summer. Along with President Lincoln, Virginians watched with interest as "Little Mac" skillfully organized the Army of the Potomac and inspired it with his personal charismatic flamboyance. Newspapers made it clear that McClellan intended to capture Richmond, the capital of the Confederacy, and put an early end to the war. With more than 100,000 Federals armed and drilled, the threat sounded far away but ominous in the sheltered Shenandoah Valley. As winter passed and spring approached, Virginians waited for McClellan to make the first move. It seemed extremely odd to everyone, and in particular to President Lincoln, when nothing happened.

A CALL TO ARMS

On March 10, 1862, Virginia governor John Letcher ordered the militia in the Shenandoah Valley to report for duty. Hotchkiss preferred to serve in the Confederate army rather than the militia, but he also wanted to be near home. He contacted a friend, Lieutenant Colonel William S. H. Baylor of Augusta County, and asked if Confederate forces in the valley could use a topographical engineer. Baylor held the post of inspector general on the staff of Major General Thomas Jonathan "Stonewall" Jackson, who commanded Confederate forces in the Shenandoah Valley District. Baylor told Hotchkiss to apply in person and provided him with letters of introduction. On March 12 Hotchkiss kissed Sara and his two daughters good-bye and headed for Staunton. Five days later, he departed for Jackson's camp with three companies of Augusta County militia.

Baylor learned that General Jackson had moved his headquarters to Hawkinstown, a small, nondescript village near Mount Jackson. Hotchkiss said he knew the way, so on March 18 he led the militia down the pike. While traveling the Valley Pike, which stretched from one end of the Shenandoah Valley to the other, the soon-to-be soldiers observed wagon trains filled with provisions and military supplies heading down the pike to Jackson's camps, as well as an unusual number of horse-drawn carts filled with personal items and people seeking safety farther up the valley. Hotchkiss had heard no reports of Federals in the lower valley and admitted to being puzzled by the sudden exodus of refugees from the area.

That evening, the column passed through Harrisonburg, and the men spent the night in a large barn. The following day, the weather turned warm, dry, and dusty. More militia joined the column as it plodded toward New Market. On March 19 the weather turned wet with snow squalls. Some of the men, Hotchkiss observed, "became soaked and chilled and suddenly fighting lost its charm." Hotchkiss remained optimistic despite the weather.

The 160th Regiment Virginia Militia made Hotchkiss their unofficial adjutant, in recognition of his service during the past summer. He realized that no position in a militia organization meant much until it was officially integrated into an army regiment, but he remained hopeful. If required, he would become a private and work his way up like the others.

"We reached camp about noon," wrote Hotchkiss, and with Major William Poague he rode to Jackson's headquarters to report the arrival of the Augusta militia. Hotchkiss had met Jackson five years earlier at the home of Reverend Doctor George Junkin, the president of Washington College in Lexington, Virginia, who was then Jackson's father-in-law. If Jackson recollected meeting Hotchkiss, he did not mention it, but he remained cordial and asked many questions about conditions in the upper valley. Nor did Hotchkiss present his credentials, as the general appeared to be heavily engaged in mustering men into the Valley Army and forming them from militia companies into regiments.

THOMAS JONATHAN JACKSON (1824–63)

BORN IN VIRGINIA ON JANUARY 24, 1824, JACKSON GREW UP IN THE HOME OF HIS UNCLE FOLLOWING THE DEATH OF HIS IMPOVERISHED PARENTS. ALTHOUGH POORLY PREPARED FOR HIGHER EDUCATION, HE OBTAINED AN APPOINTMENT TO THE U.S. MILITARY ACADEMY AT WEST POINT IN 1842. AFTER BARELY MEETING MINIMUM ACADEMIC PROFICIENCY DURING THE FIRST FEW MONTHS, JACKSON STUDIED ASSIDUOUSLY TO IMPROVE HIS GRADES. HIS CLASSMATES CALLED HIM "OLD JACK" BECAUSE HE WAS "OLDER THAN MOST MEMBERS OF HIS CLASS AND EXCEEDINGLY SHY AND BASHFUL." WHEN JACKSON GRADUATED SEVENTEENTH IN A CLASS OF FIFTY-NINE IN 1846, HIS FELLOW CLASSMATES DECLARED THAT "IF THE COURSE HAD BEEN A YEAR LONGER HE WOULD HAVE COME OUT FIRST."

AS AN ARTILLERY OFFICER, JACKSON SERVED WITH DISTINCTION IN JOHN B. MAGRUDER'S BATTERY DURING THE MEXICAN WAR AND EARNED TWO BREVETS. BORED WITH THE PEACETIME MILITARY, HE RESIGNED IN 1851 TO TEACH ARTILLERY TACTICS AND NATURAL AND EXPERIMENTAL PHILOSOPHY AT THE VIRGINIA MILITARY INSTITUTE IN LEXINGTON. TWO YEARS LATER, HE MARRIED ELINOR JUNKIN, DAUGHTER OF THE PRESIDENT OF WASHINGTON COLLEGE. ON OCTOBER 22, 1854, ELINOR DIED OF CHILDBIRTH COMPLICATIONS AFTER DELIVERING A STILLBORN DAUGHTER. THREE YEARS PASSED BEFORE JACKSON REMARRIED, THIS TIME TO MARY ANNA MORRISON, DAUGHTER OF A PRESBYTERIAN MINISTER WHO RECOGNIZED JACKSON AS BEING A TRUE AND ZEALOUS FOLLOWER, FAITHFUL TO THE DOCTRINE OF PREDESTINATION.

AS A MARRIED PROFESSOR WITH A NEW WIFE, JACKSON BECAME INCREASINGLY AUSTERE IN HIS PERSONAL HABITS. HE STOOD SIX FEET TALL, WEIGHED ABOUT 175 POUNDS, HAD BRIGHT BLUE EYES, A BROWN BEARD, UNUSUALLY ENORMOUS FEET, A LOPING GAIT, AND A CAREWORN APPEARANCE. HISTORIAN G. F. R. HENDERSON, JACKSON'S MERITORIOUS BIOGRAPHER, WROTE, "HE NEVER SMOKED, HE WAS A STRICT TEETOTALER, AND HE NEVER TOUCHED A CARD. HIS DIET, FOR REASONS OF HEALTH, WAS OF A MOST SPARING KIND." DOUGLAS SOUTHALL FREEMAN REFERRED TO JACKSON AS A MAN "OF CONTRASTS SO COMPLETE THAT HE APPEARS ONE DAY AS A PRESBYTERIAN DEACON WHO DELIGHTS IN THEOLOGICAL DISCUSSION AND, THE NEXT, A REINCARNATED JOSHUA." ALTHOUGH A GOOD ARTILLERYMAN AND A STRICT DISCIPLINARIAN, JACKSON NEVER BECAME A GOOD TEACHER. STUDENTS AT THE VIRGINIA MILITARY INSTITUTE JOKED AT HIS ECCENTRICITIES, PLAYED PRANKS ON HIM, AND CALLED HIM "TOM FOOL JACKSON." WORD SPREAD THAT HE NEVER USED PEPPER BECAUSE HE CLAIMED IT MADE HIS LEG HURT, BUT HE SUCKED LEMONS BECAUSE THEY AGREED WITH HIS DIGESTIVE SYSTEM. THE JOKES LESSENED WHEN WAR CAME. AFTER THE GENERAL'S HEROIC PERFORMANCE AT FIRST MANASSAS, HIS SOLDIERS CALLED HIM "STONEWALL." THOSE WHO HAD WATCHED HIM IN BATTLE CALLED HIM "OLD BLUE LIGHT" BECAUSE HIS EYES GLEAMED FIERCELY IN BATTLE. AT HAWKINSTOWN, THE ALREADY FAMOUS STONEWALL BRIGADE, WHICH HAD FOUGHT SO VALIANTLY AT FIRST MANASSAS, WAS RE-FORMING AS FRESH RECRUITS JOINED THE VETERANS. EVERY VOLUNTEER WANTED TO BE PART OF JACKSON'S ARMY.

THOSE SAME BLUE EYES THAT GLEAMED SO BRIGHTLY ON HENRY HOUSE HILL DURING THE BATTLE OF BULL RUN WERE GETTING READY TO SHINE AGAIN. COLONEL TURNER ASHBY, COMMANDING THE GENERAL'S CAVALRY, RETURNED TO CAMP ON MARCH 21 TO REPORT HEAVY SKIRMISHING WITH PART OF MAJOR GENERAL NATHANIEL P. BANKS'S FEDERALS AT KERNSTOWN, FORESHADOWING HARD FIGHTING IN THE WEEKS AHEAD. ASHBY ALSO BELIEVED THAT PART OF MAJOR GENERAL JAMES SHIELDS'S DIVISION OF BANKS'S FIFTH CORPS WAS BOARDING TRAINS AND BEING DETACHED FROM THE VALLEY IN READINESS FOR SERVICE ELSEWHERE.

THE EVE OF BATTLE

Hotchkiss watched Jackson as the general moved through the camps of militia companies. He instinctively admired the thirty-eight-year-old officer without knowing why, and his opinion never changed. Without realizing it, Hotchkiss and Jackson were in some respects alike. They were both deeply religious, placed duty above all else, and performed their responsibilities without any expectation of praise. In other respects they were much different. Jackson never wasted words, being straight to the point; Hotchkiss rarely kept quiet because he enjoyed talking.

Hotchkiss waited every day for an opportunity to present his letters of recommendation to the general but could not bring himself to do it. On March 21 Colonel Baylor inducted part of the Augusta militia into the regular army for the duration of the war, unless discharged sooner. Three hundred and eighty men were not enough for a regiment, so Major Poague formed a battalion. Hotchkiss retained his post as adjutant, at least until the battalion became part of a regiment. "Col. Baylor says Gen. Jackson will have me detailed for engineer duty," Hotchkiss noted in his journal, but no affirmation came from the general.

During the day, Jackson began to move his army down the Valley Pike toward Winchester, the principal city in the lower valley where major crossroads linked the town with eastern and western Virginia. Rumors spread and men talked. Somewhere in the lower valley there would be fighting. On the night of March 22, the men of the Valley Army stayed up late writing to loved ones—and for some, those letters would be their last.

THE BATTLE OF KERNSTOWN: MARCH 23, 1862

On March 12 Major General Nathaniel P. Banks sent General Shields's division to Winchester. Shields's Federals drove off Jackson's skirmishers and took possession of the town. Banks intended to occupy and control the entire valley, and his first objective was to secure a Federal hold on Winchester. His orders called for the destruction of Jackson's Valley Army, which was a small division detached from General Joseph E. Johnston's Department of Northern Virginia. The Union war department also tasked Banks with preventing any raids into Maryland or allowing any Confederate military unit to threaten Washington during General McClellan's Peninsula campaign. Believing he had secured the valley, Banks began to shuttle a few units eastward to reinforce McClellan's Army of the Potomac. On March 21 Ashby's Confederate scouts observed Federals boarding trains and informed Jackson that Shields's force at Winchester was being reduced. One of Jackson's objectives had been to hold Banks's corps in the valley and prevent him from reinforcing McClellan. Taking the majority of his 4,500 volunteers from camps in the vicinity of Mount Jackson, Stonewall began a series of forced marches with two-thirds of his command to remind Banks that the Valley Army intended to fight.

Ashby's information led Jackson to believe that most of Shields's division had withdrawn, leaving only a small rear guard of about four infantry regiments at Winchester. Had Ashby studied the Federal activity more thoroughly, he would have observed that Shields still occupied Winchester with a division of 9,000 men.

Jackson arrived on March 23 at Kernstown, three miles south of Winchester, with a force of about 3,000 infantry weary from marching thirty-six miles in thirty-six hours. Although Jackson disliked fighting on Sundays, he struck with his entire force without first reconnoitering.

After a quick feint on level ground near the pike, Jackson's main force of three undersized brigades moved into the woods northwest of town. The Confederates charged with a small brigade under Colonel Jesse S. Burks to gain possession of the Valley Pike while two slightly larger brigades under Colonel Samuel V. Fulkerson and Brigadier General Richard B. Garnett moved to the ridge on Pritchard's Hill to flank the enemy. At a stone fence running along the ridge, Fulkerson and Garnett met stiff resistance from two Federal brigades commanded by colonels Erastus B. Tyler and Nathan Kimball. Farther to the right and east of the pike, a diversionary attack by Ashby's cavalry struck Colonel Jeremiah C. Sullivan's brigade. The cavalry charge encountered heavy fire and fell back in disorder. Jackson did not have a commanding view of the battlefield. Expecting only light resistance, he remained east of Kernstown with his artillery and a few companies of infantry that wasted time dueling with Federal skirmishers.

On the far right, Fulkerson, Garnett, and Burks fought a bitter struggle against Kimball and Tyler for possession of a stone fence. The unexpected strength of Shields's division surprised Jackson and caused turmoil among his brigade commanders. Because Jackson had kept his plans to himself and was not at the ridge, he could not understand why his brigades were being repulsed. He grew concerned when Garnett's Stonewall Brigade ran out of ammunition and called for reinforcements. Jackson had no reinforcements, having committed every unit on the field to driving what he believed to be a few Federal regiments out of Winchester. Compelled to retire for want of ammunition, Garnett's withdrawal forced Fulkerson and Burks to follow. Furious by the withdrawal of the Stonewall Brigade, Jackson relieved Garnett. Finally admitting he had been beaten, Jackson saved most of his artillery and wagons but posted 700 casualties. He collected his wounded and moved back about four miles to Newton. Ashby checked an uninspired Federal pursuit a mile south of Kernstown, and the battle ended ignominiously for the Valley Army.

Although Jackson lost the battle of Kernstown, he accomplished his other mission. General Banks ordered the troops recently transported away from Winchester back to the valley, and because Jackson had unknowingly attacked a force of 9,000 Federals, Washington overestimated the size of the Valley Army and made strategic errors by depriving McClellan of troops for the Peninsula campaign.

BATTLE OF KERNSTOWN CASUALTIES

	KILLED	WOUNDED	MISSING	TOTAL
CONFEDERATE	80	375	263	718
FEDERAL	118	450	22	590

Battle of Kernstown [Map #91]

Hotchkiss did not participate in the battle of Kernstown because his militia battalion had not been organized into a Confederate regiment. Later, after he began to map the Shenandoah Valley, he took special interest in Kernstown and, with assistance of officers who fought there, produced an after-action map. Hotchkiss drew many battle maps in this manner because he had developed sketches and used them to produce maps. He carefully detailed the Kernstown map, which became the model for many of the battle maps he later composed.

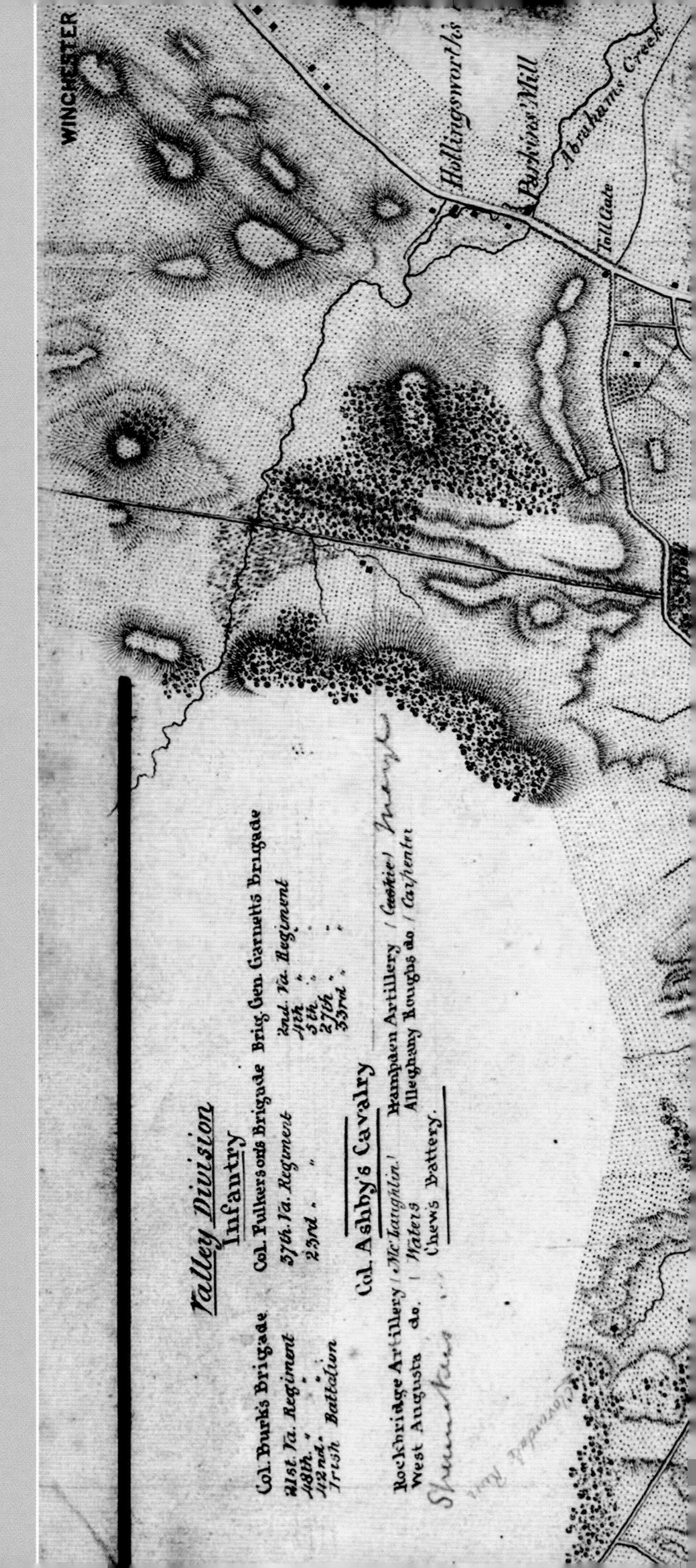

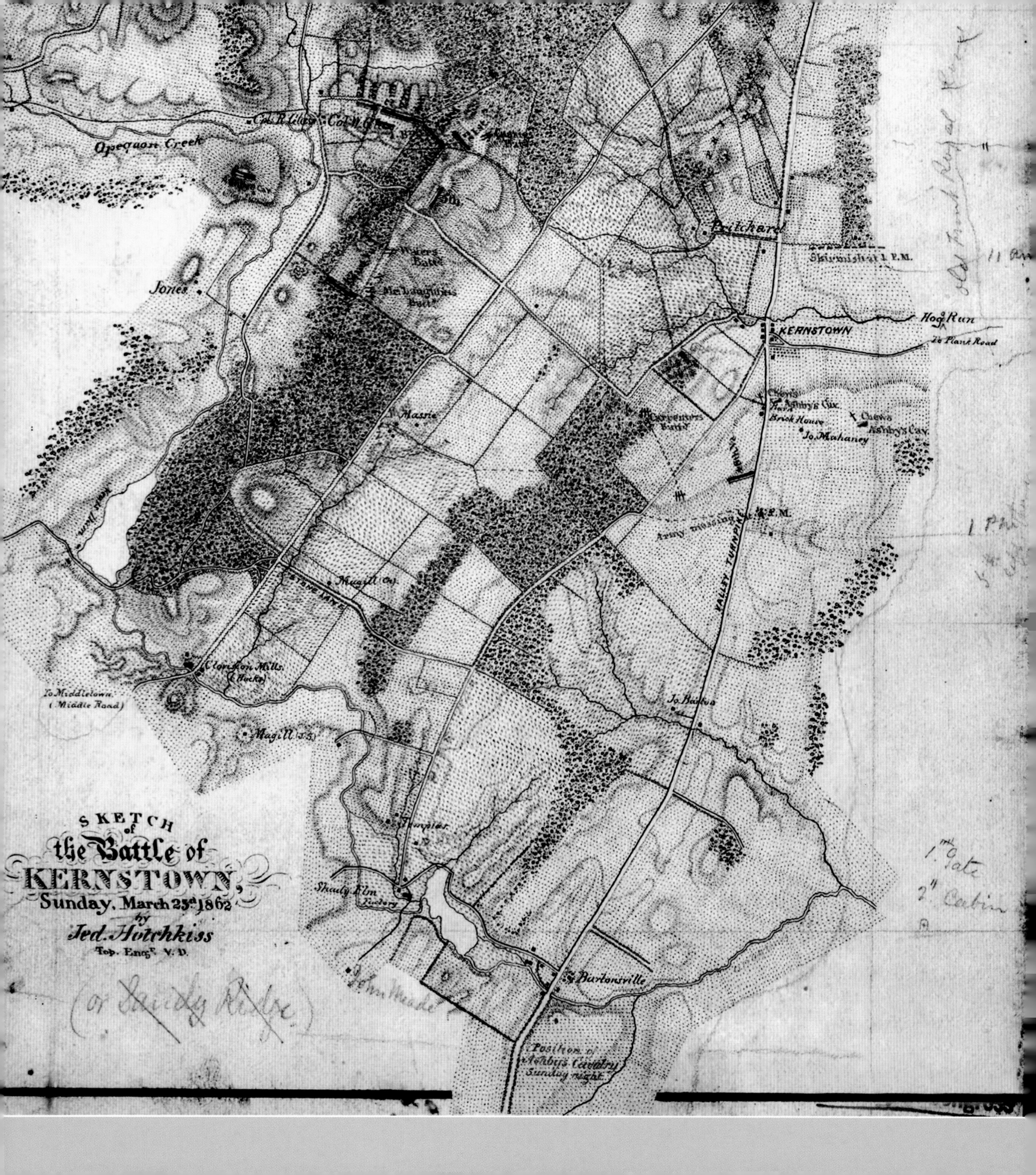
SKETCH of the Battle of KERNSTOWN, Sunday, March 23rd 1862 by Jed. Hotchkiss Top. Engr. V. D.
KERNSTOWN
VALLEY TURNPIKE
Hog Run
To Plank Road
Opequon Creek
Jones
Pritchard
Skirmish at 1 P.M.
Kernstown
W. Massie
Magill
Stone Lane
Carpenter's Batt.
Ashby's Cav.
Brick House
Jo. Mahaney
Jo. Barton
Glass
Barton's Mills
To Middletown (Middle Road)
Shady Elm
Bartonsville
Position of Ashby's Cavalry Sunday night

STONEWALL'S TOPOGRAPHER

Hotchkiss regretted not being on the field to witness the battle of Kernstown. Many of the men from Augusta County had already joined veteran regiments, fought at Kernstown, and some of them did not come back. Major Poague wanted to form a regiment with 380 men, but the war department specified 1,000-man regiments. Despite being mustered into the Confederate service, Hotchkiss remained a part of Poague's Augusta County Militia Battalion. The unit had not been armed, and Hotchkiss remained behind at Mount Jackson, a two-day march to Kernstown.

Late in the day, a courier arrived with orders for the battalion to draw weapons and march north to rendezvous with Jackson's retreating army. On March 24, as the battalion hiked along the turnpike, a stream of ambulances filled with Kernstown's wounded passed. Poague halted the battalion near Woodstock when he encountered the general riding by himself and looking gloomy. Jackson ordered the battalion to fall back to Woodstock, which some of the men interpreted to mean the Federals were in pursuit, making them nervous. Poague and Hotchkiss calmed the men and turned them about in good order. Jackson watched from the roadside, observing Hotchkiss's self-composure.

Road-weary and thoroughly exhausted, the survivors from Kernstown paused to spend the night at Woodstock before going into camp at Narrow Passage. During the retreat, Hotchkiss had kept the column closed up and ensured that the stragglers remained in formation. After settling into camp that night, he received two letters from Sara, who warned that their daughter Nelly had been smitten with scarlet fever. The following day another letter arrived: Nellie was still alive but fighting for her life.

On the morning of March 26, Jackson ordered Hotchkiss to report to headquarters, a private home near Narrow Passage Creek. The general wasted no time getting to the point. After questioning Hotchkiss about his topographical work in western Virginia in 1861, Jackson said, "I want you to make me a map of the Valley, from Harpers Ferry to Lexington, showing all the points of offence and defence in those places. Mr. [Alexander S.] Pendleton will give you orders for whatever outfit you want. Good morning, sir." Hotchkiss asked Pendleton—Jackson's adjutant general—for a wagon, two horses, a driver, a tent, and camp equipment. Though still very much concerned about his daughter's illness, he bid good-bye to his battalion and rode off to begin, in his own words, the "big job."

THE "BIG JOB"

Mapping the Shenandoah Valley was not just an immense undertaking because it was big. At 140 miles in length and up to twenty-five miles in width, it was massive. The actual valley lay between two parallel mountain ranges, with North Mountain and the Alleghenies to the west and the Blue Ridge Mountains to the east. As an added complication, the narrow, fifty-mile-long Massanutten Mountain ran from Strasburg to Harrisonburg and between the two ranges. Made of sandstone, the Massanutten split the flow of the Shenandoah River into the North Fork and the larger South Fork, which came together at Front Royal. In addition to rivers, tributaries, and mountain gaps, there were few good roads but hundreds of backcountry byways winding through hills, meadows, and farmlands. Organizing a map to include every feature would seem an impossible task to almost anyone but Hotchkiss, who thoroughly enjoyed untangling the labyrinth of terrain.

On March 28 Hotchkiss began his first survey on behalf of the general by studying the area in the vicinity of the Confederate encampment. Following Narrow Passage Creek in the direction of North Mountain, Hotchkiss made an important discovery. Jackson believed he had placed his force in a strong defensible position, but Hotchkiss discovered several roads that could be used by the enemy to flank the Valley Army. He returned to camp and showed the general the risks of remaining where they were. That afternoon Jackson issued orders to fall back to Pence's Hill. He then made another decision because of Governor Letcher's order to draft all militia as privates into volunteer companies. Jackson moved quickly, named Hotchkiss acting topographical engineer, and made him a civilian employee with lieutenant's pay on the headquarters staff.

Not entirely satisfied with Pence's Hill, Hotchkiss rode out of camp the following morning on a reconnaissance with Ashby's cavalry and a local guide familiar with the Woodstock region. After sketching the area's roads, streams, farms, and obstacles, Hotchkiss returned to headquarters and informed Jackson that there were no defensible positions north of Woodstock. By March 31, Hotchkiss had worked into the area around Edenburg and found an excellent defensive position at Stony Creek. Union cavalry had been active in the area, so Hotchkiss made a sketch and carried it back to headquarters. He advised Jackson to move the army and make a stand at Stony Creek, which was deep and wide. Woods half a mile to the south provided perfect cover for Confederate snipers to harass the enemy. The general agreed and sent Hotchkiss back to Edenburg with infantry and a message for Ashby to hold the line. Hotchkiss came under fire from Federal sharpshooters at Edenburg, but Ashby's cavalry drove them off and burned the bridge over Stony Creek. A series of skirmishes ensued, but the line held and Jackson reestablished headquarters a few miles away at Rude's Hill. (For the map of the Strasburg–Mount Jackson area, see page 39.)

Hotchkiss was only one of many who witnessed the recruiting process as he marched through the towns of the Shenandoah Valley on his way to Jackson's camp.

WITHDRAWING UP THE VALLEY

Although defensively situated on the south side of Stony Creek, Jackson's main mission of keeping the Federals engaged required constant sorties against the enemy. He sent Ashby's cavalry and a brigade of infantry to slow the advance of Shields's division. Jackson could not risk exposing his three small brigades to the overwhelming strength of the Federals, so he continued his strategy of grudgingly withdrawing up the valley. Whenever Shields stalled, Jackson rested his veterans and drilled the new regiments in an effort to prepare them for the fighting ahead. He used every hour Banks allowed him to train and strengthen the Valley Army.

During the withdrawal, Hotchkiss spent his days riding through the country, making sketches, determining distances, following roads to see where they led, and locating defensive sites where the army could camp with less fear of being surprised. Jackson soon realized that without Hotchkiss's relentless scouting and observations, the army could be in constant jeopardy. Taking a road that turned to mud in a rainstorm could be disastrous when the surface of another nearby road shed water quickly and remained hard. Hotchkiss even picked campsites, avoiding woodlands filled with soft pine and instead choosing areas with oak because hardwoods burned clean without raising clouds of black smoke. He also evaluated the condition of crops to determine whether they could be used as fodder for the animals or food for the men. Although Hotchkiss often worked alone, Jackson grew concerned about the possibility of his mapmaker being captured by enemy scouts and began to provide him with a handful of scouts from Ashby's cavalry.

Hotchkiss also obtained help. Sergeant S. Howell Brown had surveyed part of Jefferson County, and Jackson agreed to transfer him to Hotchkiss's topographical staff. While Hotchkiss worked on the upper valley, Brown worked on the lower valley. Brown eventually became a disciple of Hotchkiss's topographical techniques and produced two outstanding maps for the atlas, one covering the Maryland expedition during the summer of 1862 and the other the Gettysburg campaign in 1863 (see page 105, Map #43-1).

REBUILDING THE ARMY

Reinforcements arrived in mid-April. General Joseph E. Johnston, commanding in Virginia, sent Brigadier General William B. Taliaferro to assume command of the Third Brigade, which Fulkerson had ably led. Jackson protested but lost the argument. With Colonel Burks away on extended sick leave, Jackson pulled Colonel John Campbell from the Forty-eighth Virginia and put him in charge of the Second Brigade. The Stonewall Brigade created a row when Brigadier Charles S. Winder, a West Pointer from South Carolina, arrived to replace Garnett, whom the general removed from command after Kernstown. Jackson could not have asked for a better brigade commander than Winder, who responded to a chilly reception from malcontents by demanding discipline and leading the brigade through some of the army's most grueling campaigns.

Jackson knew he could not accomplish much without more infantry, and asked for 17,000 men. "I could so threaten the enemy's rear as to induce him to fall back and thus enable me to attack him whilst retreating . . . If Banks is defeated it may greatly retard General McClellan's movements." Jackson suspected that Johnston did not have 17,000 men to spare, but he promised that with those sent, "no stone shall be left unturned to give us success." No reinforcements arrived, but in mid-April, after General Johnston abandoned the Confederate works at Manassas, he moved Major General Richard S. Ewell's division to the Rapidan River and authorized Jackson to bring it to the Shenandoah, if needed.

Jackson immediately began to correspond with Ewell over the topographical complexities of the Shenandoah Valley. He did not detach Hotchkiss to discuss geographic problems with Ewell because on April 17 Ashby's cavalry suffered a defeat at Stony Creek and fell back to Rude's Hill. With the enemy pressing in force, Jackson expected his weak command to be pushed farther up the valley and asked Hotchkiss to lay out the terrain. He also discovered that Brigadier General Louis Blenker's division from Major General John C. Frémont's Mountain Department had joined Banks. With only 5,000 men in his command and with 30,000 Federals pressing against his front, Jackson summoned Ewell to Swift Run Gap, positioned him on the southeastern tip of Massanutten Mountain, and told him to wait there. As Ewell moved toward the valley, Jackson pulled back to Harrisonburg on the southwestern tip of Massanutten Mountain, directly opposite from Ewell.

What Ashby's cavalry lacked in discipline they made up for in gallantry. The combination did not always work, and when Jackson decided to withdraw to Harrisonburg, he pulled Hotchkiss away from mapmaking and sent him on an errand. To keep Banks out of the Luray Valley, he asked Hotchkiss to find Ashby's cavalry and order them to burn all the bridges. Hotchkiss had been mapping the area around Massanutten Mountain and knew where the bridges were located, but it took time to track down Ashby's cavalry. He finally located two companies, one under Captain George Sheetz and the other under Captain Macon Jordan, all guzzling applejack in the Shenandoah Iron Works.

On Hotchkiss's orders, the cavalry saddled up and galloped off in an intoxicated hoopla to burn the Red, Columbia, and White House bridges across the South Fork of the Shenandoah. The three bridges carried roads leading to New Market. Hotchkiss split the two cavalry companies into three detachments and sent each off to destroy a bridge. At the same time, Federal cavalry, infantry, and artillery appeared on roads leading to New Market to secure the spans. Sheetz and Jordan fled, and many of the drunken soldiers disappeared over the Blue Ridge and were not seen for several days. Hotchkiss rallied the small force he guided and at the first opportunity burned the Red Bridge. The cavalry's failure to destroy the other two bridges forced Jackson to retire farther up the valley.

Burning the Shenandoah's Bridges
During the Valley Army's withdrawal in April 1862, Hotchkiss prepared a hurried sketch of roads and of the three bridges in the Luray Valley—White House Bridge, Columbia Bridge, and Red Bridge—that Jackson ordered Ashby's cavalry to burn.

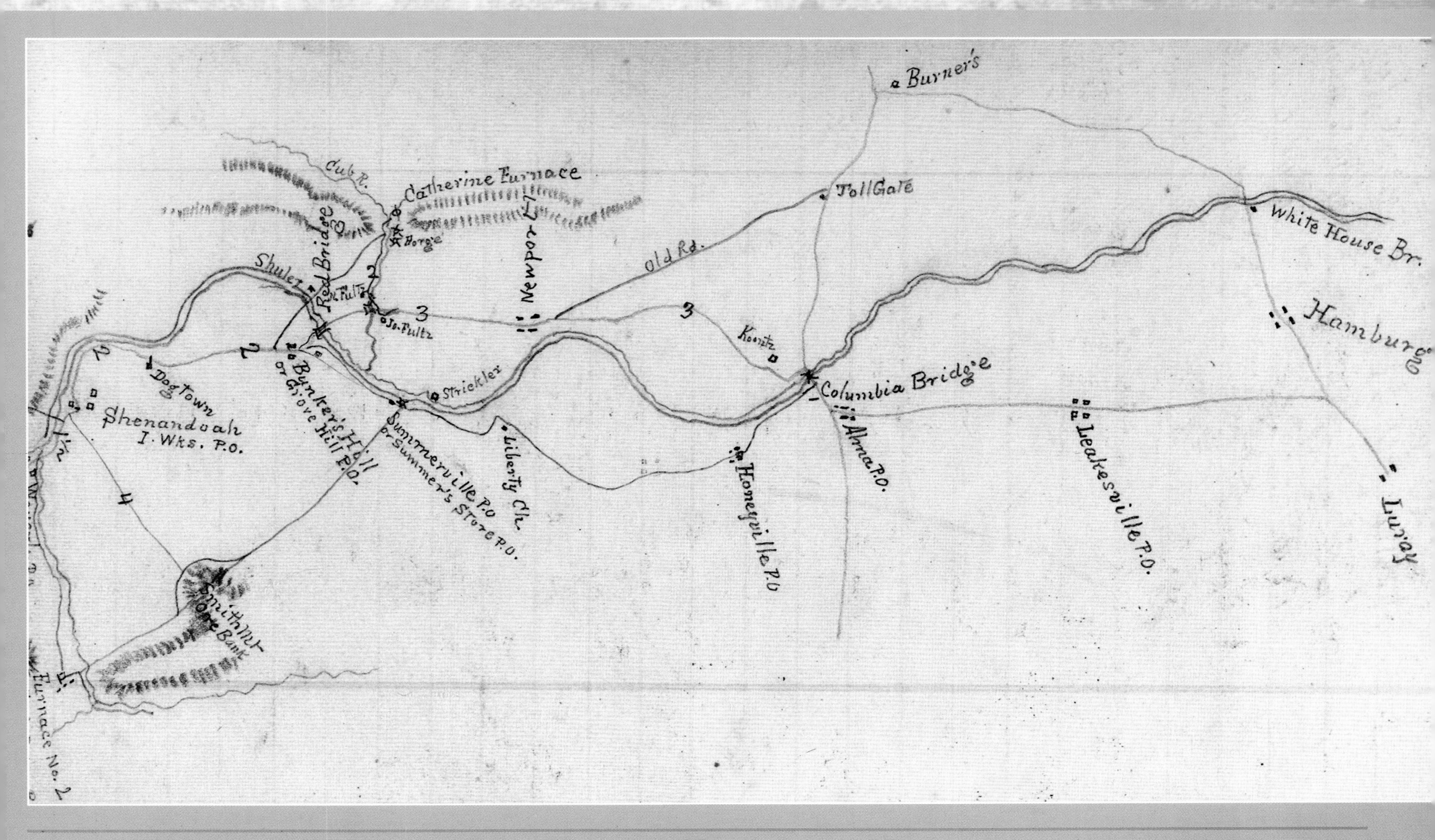

TURNER ASHBY (1828–62)

When Virginia seceded in the spring of 1861, Ashby joined Jackson's forces at Harpers Ferry and eventually became a lieutenant colonel in Brigadier General James E. B. Stuart's Confederate cavalry. Ashby preferred serving in the valley and in October 1861 became Jackson's commander and colonel of the Seventh Virginia Cavalry Regiment.

During the spring of 1862 and after the battle of Kernstown, Ashby moved up to brigade command in the Valley Army. Completely fearless and reckless in battle, Ashby performed brilliantly during operations in the valley, but he operated an undisciplined command that sometimes cost Jackson lost opportunities. Because of Ashby's personal bravery, his men too often lost their lives attempting to emulate him. He constantly irritated Jackson because the cavalry enjoyed working independently, but they were great scouts and indispensable during the Shenandoah Valley campaign. On June 6, 1862, Ashby lost his life as a result of friendly fire near Harrisonburg, Virginia, while performing a rearguard operation two weeks after being promoted to brigadier general. He is buried in Stonewall Cemetery in Winchester, Virginia.

Ashby grew up on a farm in Fauquier County in the Shenandoah Valley and received much of his education at home from his mother and tutors. He became a grain dealer, planter, and local politician of wealth. Ashby had never received military training, but he became a superb horseman. He formed a cavalry company of volunteers and in 1859 led it to Harpers Ferry after abolitionist John Brown attacked the federal armory.

"When Ashby's men are with him they behave gallantly, but when they are away they lack the inspiration of his presence, and being undisciplined, they often fail to do any good."

—FROM HOTCHKISS'S DIARY, APRIL 19, 1862

Although Ashby had not been directly involved in the abortive operation, Jackson stripped him of command. Ashby located Hotchkiss and railed against being deposed by the general. After reaching an emotional climax and against Hotchkiss's advice, Ashby sent his resignation to General Winder, to whom he had been reassigned. Winder understood that Jackson could not function without Ashby and throughout the night mediated the issue between the romantic cavalier of storybook legend and the stubborn, pragmatic general who believed every soldier must explicitly follow orders. Jackson soon realized that if he accepted Ashby's resignation, the entire cavalry would revolt, resign, reorganize, and follow their leader in an independent command. With thousands of Federals pouring up the valley, Jackson decided a disorganized cavalry under Ashby was better than none and thus rescinded his order. Ashby withdrew his resignation and resumed virtually unfettered command on April 26.

THE SITUATION IN VIRGINIA

When in early April General McClellan put the 112,000-man Army of the Potomac on Virginia's peninsula, and Major General Irvin McDowell moved a 35,000-man Army of the Rappahannock into the Fredericksburg area, General Johnston considered recalling Ewell's division for the defense of Richmond. General Lee, who was then serving on President Davis's staff, thought it might be wiser to let Jackson retain Ewell's division to engage Banks.

After a lengthy exchange of correspondence on other options, Lee left the decision to Jackson, writing, "The blow, wherever struck, must, to be successful, be sudden and heavy. The troops used must be efficient and light." Jackson deliberated. Johnston had ordered him to fight on the defensive, but it now appeared that Lee wanted him to take the offensive. Lee also authorized Jackson to strike beyond the valley, an option Johnston had never sanctioned. After rereading Lee's instructions, Jackson decided he would strike and elevate the Valley Army from a defensive sideshow to the main attraction.

On the eve of the Shenandoah Valley campaign, Hotchkiss noticed the determination in the eyes of the men serving in the Stonewall Brigade. Jackson, an old artilleryman himself, took special interest in preparing the men serving his batteries for the forthcoming campaign.

CHAPTER TWO: STONEWALL JACKSON'S VALLEY CAMPAIGN

CRISIS IN THE VALLEY

With freedom of action, Jackson went first to his maps. By the end of April, General Banks's 20,000-man corps had entered Harrisonburg and pushed the Valley Army about as far as Jackson intended to be pushed. In addition to General Frémont's 20,000-man army in western Virginia, Banks could draw another 5,000 men from outposts and consolidate a force of 45,000 men. Frémont's vanguard of 3,700 men under Brigadier General Robert H. Milroy had already pushed Brigadier General Edward "Old Allegheny" Johnson's Army of the Northwest, which was actually a brigade, toward Staunton. Frémont intended to strike from the upper valley while Banks struck from the lower valley, and together they would mash Jackson's army in the middle.

Jackson also had a plan. He intended to strengthen his own force with Johnson's beleaguered brigade, prevent the Federals from combining by keeping Banks stationary and defeating Milroy, and protecting Staunton's railroad link, through which the Virginia Central passed. Jackson began to consolidate his forces in late April. By adding Johnson's 2,500-man brigade and Ewell's 8,000-man division, the Valley Army swelled to about 18,000 men. How Jackson planned to use this force depended profoundly on the maps that had been assembled by Hotchkiss over the past six weeks.

Obsessed with secrecy, Jackson refused to discuss his plans with anyone, but in late April he sent Hotchkiss into western Virginia with maps to confer with Johnson. There had been little direct communication between Jackson and Johnson, but Hotchkiss suspected his trip was for the purpose of hatching a military movement. East of the Blue Ridge at Swift Run Gap, Ewell grew increasingly impatient because Jackson refused to include him in the planning. In late April, Ewell decided that his superior was insane and could not be trusted.

On April 30 Jackson began to move his army away from Swift Run Gap and eastward through Brown's Gap. Ewell began to move his division into Swift Run Gap and occupied the camp abandoned by Jackson. Hotchkiss remained that day with Ewell and spent several hours on Peaked Mountain, the southernmost crest of the Massanuttens, observing the movements of Banks. When it appeared that Banks intended no further movements up the valley, Hotchkiss conveyed the intelligence to Jackson by standing on the peak and signaling with a bedsheet.

By May 3, more than 6,000 men from the Valley Army had passed through Brown's Gap and appeared to be on their way to Richmond. Every inhabitant from Port Republic to Harrisonburg believed Jackson had abandoned the valley. Banks advised Washington that he had driven Jackson from the Shenandoah, and considered his mission complete.

For several days, the Valley Army repaired the tracks of the Virginia Central, which ran from Charlottesville to Staunton. Jackson halted his force at Mechum's River Station and on the morning of May 4 began one of the most brilliant strategic campaigns in military history.

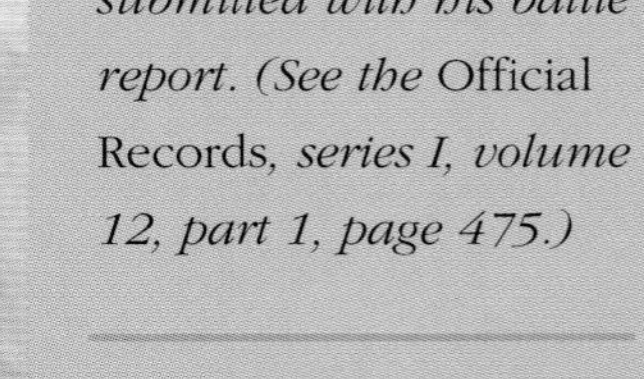

Battle of McDowell [Map #93]
Hotchkiss assembled several small sketches made in the field and pieced them together a short time later to produce a composite sketch of the battlefield. He later added more detail to the map General Jackson submitted with his battle report. (See the Official Records*, series I, volume 12, part 1, page 475.)*

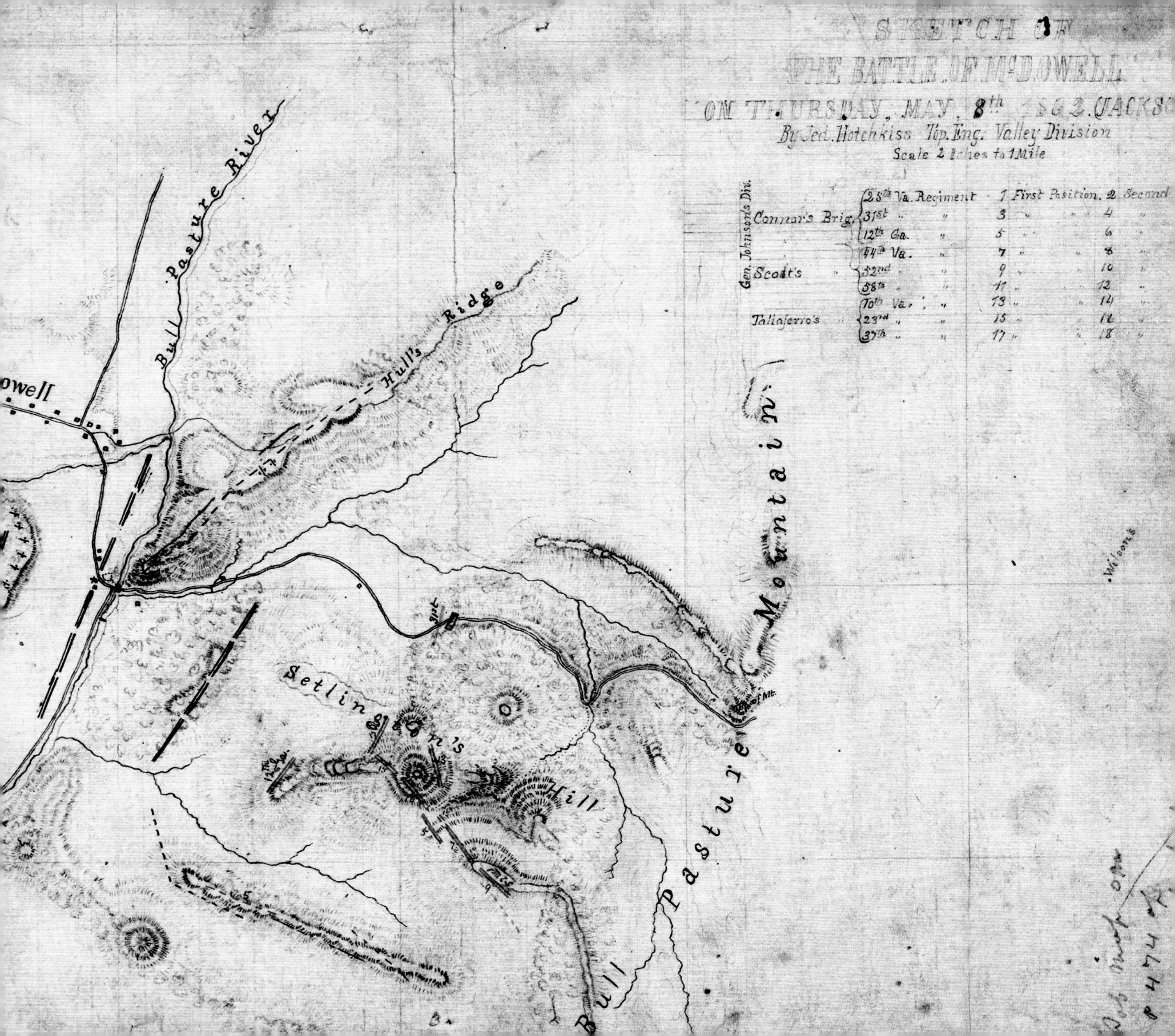
SKETCH OF
THE BATTLE OF McDOWELL
ON THURSDAY, MAY 8th 1862. JACKSO
By Jed. Hotchkiss Top. Eng. Valley Division
Scale 2 Inches to 1 Mile
Gen. Johnson's Div.
Connor's Brig.
25th Va. Regiment 1 First Position. 2 Second
31st 3 4
12th Ga. 5 6
44th Va. 7 8
Scott's
52nd 9 10
58th 11 12
Taliaferro's
10th Va. 13 14
23rd 15 16
37th 17 18
Bull Pasture River
Hull's Ridge
owell
Setlington's Hill
Bull Pasture Mountain
Wilson's
Bull Pasture

In late April, General Jackson learned from Ashby's scouts that General Frémont had started moving his 20,000-man Federal army toward Staunton to cooperate with General Banks's 20,000-man corps at Harrisonburg. General Milroy's 3,500-man brigade led Frémont's approach on roads passing through McDowell, a village about thirty miles northwest of Staunton.

On April 30, when the Valley Army began to move east of the Blue Ridge Mountains, Jackson conceived a complicated strategy he shared with no one but Hotchkiss, who prepared the routes. Jackson communicated with General Johnson only once and told him to move his brigade fourteen miles to West View, which was located about halfway between Staunton and McDowell. After marching the Valley Army—all but Ashby's cavalry and Ewell's division—through Brown's Gap and to Mechum's River Station, Jackson stopped. He gave the troops a brief rest while loading artillery and equipment on railroad cars. Instead of proceeding to Richmond as Banks expected, Jackson reversed his course, sent the wagons back through Rockfish Gap, and on May 6 marched into Staunton. By then, Jackson's "foot cavalry" had covered ninety-two miles in four days, and another twenty miles from Mechum's River Station to Staunton.

After assimilating Johnson's brigade at West View, Jackson now led a combined force of about 10,000 men opposed to Milroy's 3,500 and Brigadier General Robert C. Schenck's 3,300. Jackson detailed Hotchkiss to lead the advance over a series of winding mountain roads on which the latter had driven during the Rich Mountain campaign. Late in the afternoon on May 7, Johnson's skirmishers drove Milroy's pickets back without forcing an engagement. Milroy immediately sent for Schenck's brigade, doubling the number of blueclads on the ground.

On the afternoon of May 8, Jackson moved his division on Sitlington's Hill, which overlooked Federal dispositions on the far side of the Bull Pasture River. By then, most of Jackson's men had marched another thirty-four miles in twenty-three hours. Meanwhile, Schenck arrived with his brigade and by virtue of seniority took command of Milroy's brigade. While Hotchkiss and Ashby's scouts roamed through farm roads and logging trails looking for a way to flank the Federals, Schenck attacked.

Jackson's dispositions on Sitlington's Hill were poorly organized. Confederates on the crest had to stand to fire, and their outlines stood in bold relief against the sky. Confederate accuracy also suffered. When firing from an elevated position, bullets tended to go high. The action eventually became fierce, mainly because Schenck's Ohio and western Virginia regiments fought with the same characteristic vigor as Southern troops. Jackson brought up another brigade with the intention of counterattacking, but at dusk, after four hours of fighting, the Federals fell back to the Bull Pasture River. Intending to resume the fight in the morning, Jackson therefore rearranged his lines, but during the night Schenck withdrew into the Allegheny Mountains with Ashby's cavalry in pursuit.

The Valley Army pursued in the morning, but the roads were narrow, rutted, and strewn with debris. Schenck conducted effective rearguard actions and set fire to a stand of woods that blew hot ashes and thick, stinging smoke into the faces of the Confederates. On May 12 Jackson called a halt. He left part of Ashby's cavalry behind to annoy Schenck and began marching back to the valley to undertake the next phase of the campaign.

The casualties did not tell the full story. Jackson lost one of his best generals when Johnson suffered a wound that kept him out of the war for a year.

BATTLE OF McDOWELL CASUALTIES

	KILLED	WOUNDED	MISSING	TOTAL
CONFEDERATE	45	423	0	468
FEDERAL	26	227	3	256

HOTCHKISS AT McDOWELL

Much of Jackson's success at McDowell resulted from Hotchkiss's role in laying out the terrain and locating positions of advantage. Hotchkiss led the advance because he knew the roads. At every bend in the terrain, he stopped to study the way ahead and signaled Jackson with a handkerchief if the winding roadways were clear. When he reached Bull Pasture Mountain, which lay behind Sitlington's Hill, he took Jackson to the crest, from which the Federal line could be observed from one end to the other. While there, he sketched a map on the spot so the general could issue clear instructions to his officers. When Jackson wanted his artillery positioned, he detailed Hotchkiss to find a pathway over which the guns could be transported through woods and placed in front of Sitlington's Hill. During the afternoon of May 8, when Jackson heard fighting on the hill, he ordered Hotchkiss to ride over and tell General Taliaferro to hold on because reinforcements were on the way. Hotchkiss found the battlefield already strewn with bodies. As night approached, the fighting subsided, but not before Hotchkiss prepared some sketches so Jackson could direct the fighting in the morning. During the night, Schenck abandoned McDowell and escaped into western Virginia.

On May 10 Jackson began the pursuit, but only to drive Schenck farther from Banks. He wanted to find a way to block Schenck and keep him from turning back toward the valley, and asked his topographer for advice. Hotchkiss replied that blocking the North and Dry River gaps would detain Schenck. Jackson told Hotchkiss to do it, but to keep him informed every hour. Hotchkiss located Captain Franklin F. Sterrett's company of Ashby's cavalry and led them into the gaps. The troopers spent the day felling trees and rolling boulders down hillsides and into the roadways, completely obstructing both gaps. With Frémont's two brigades still retreating into western Virginia, Hotchkiss rejoined Jackson and started back to Ewell's camp at Swift Run Gap.

From the beginning to the end of the McDowell campaign, Hotchkiss may have been the only other person to know the general's plans. Jackson would pull the mapmaker aside, completely out of range of anyone else's hearing, and explain exactly what he intended to accomplish each step of the way, and what he needed from Hotchkiss to make his strategy work. No one would be more involved in Jackson's next move than Hotchkiss.

On May 8, as Jackson prepared to engage Schenck and Milroy at McDowell, Banks received orders from Washington that he should detach 11,000 men and send them east to General McDowell's Army of the Rappahannock. Banks complied with this and detached General Shields's division. Although Banks still believed that Jackson had taken the Valley Army to Richmond, he felt insecure after detaching half his force and began to start a slow withdrawal from Harrisonburg to Strasburg.

Jackson returned to Swift Run Gap, conferred with Ewell, and began to formulate a plan to pursue and attack Banks. Jackson had to first get his army over the swollen North River and so turned to his topographer for suggestions. Hotchkiss had lived for

Brigadier General William Booth Taliaferro (1822–98) *commanded a brigade under Stonewall Jackson during the Shenandoah Valley campaign and eventually took command of Jackson's division at Cedar Mountain. Hotchkiss worked closely with Taliaferro and a warm friendship developed.*

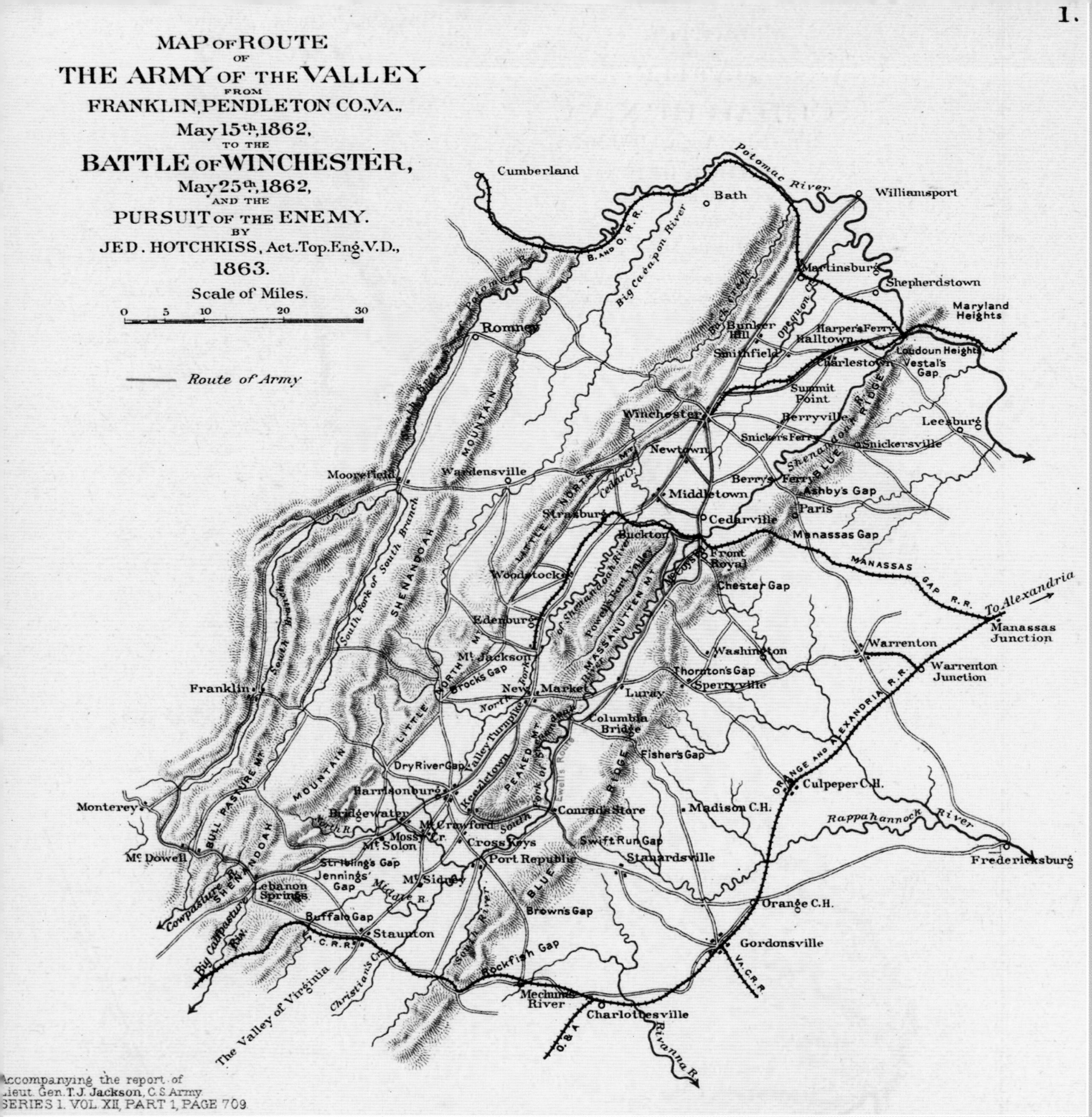

MAP OF ROUTE
OF
THE ARMY OF THE VALLEY
FROM
FRANKLIN, PENDLETON CO. VA.,
May 15th, 1862,
TO THE
BATTLE OF WINCHESTER,
May 25th, 1862,
AND THE
PURSUIT OF THE ENEMY.
BY
JED. HOTCHKISS, Act. Top. Eng. V.D.,
1863.

Scale of Miles.
0 5 10 20 30

Route of Army

Accompanying the report of
Lieut. Gen. T.J. Jackson, C.S. Army.
SERIES 1. VOL XII, PART 1, PAGE 709.

Movements in the Valley [Map #85-1]

Hotchkiss produced an after-action map of the first phase of the Shenandoah Valley campaign. Because he drew it to accompany General Jackson's battle report, he titled it the "Map of Route of the Army of the Valley." The map shows Jackson's actual movements from May 15 to May 25, 1862.

many years at nearby Mossy Creek and had attended church and social gatherings with many of the local farmers. Most of them owned durable four- and six-horse wagons, which, Hotchkiss suggested, could be moved into the river as trestles and planked over like pontoon bridges. Jackson liked the idea. Hotchkiss rounded up the wagons, and on May 18 the Valley Army crossed near Mount Solon.

The next morning, Jackson began to move down the valley using the pike on the west side of Massanutten Mountain while Ewell started to move north on a parallel course using roads on the other side of the mountain. Banks soon learned that Jackson, after defeating Schenck, had returned to the valley and was now marching down the pike in pursuit.

THE BATTLE OF FRONT ROYAL: MAY 23, 1862

At New Market, Jackson suddenly turned east and sent the Valley Army through a gap in the Massanuttens to join Ewell's division in the Luray valley. The sudden disappearance of Jackson was baffling to Banks. Suspecting that the Valley Army had crossed through one of the gaps, Banks sent about 1,000 men and a pair of guns to Front Royal, which lay east of Strasburg and near the northern end of Massanutten.

With a force of 17,000, Jackson's foot cavalry moved swiftly, maintaining a rapid pace for fifty minutes every hour. After a ten-minute rest, the men were back on the road, moving north in unison and following the South Fork of the Shenandoah River through one of the most delightful and scenic regions in the valley.

Jackson stopped to reconnoiter the road running between Strasburg and Front Royal and sent Hotchkiss ahead with Lieutenant James K. Boswell, the Valley Army's chief engineer. Hotchkiss made sketches of the area while Boswell climbed Signal Knob on the north end of Massanutten Mountain. The two scouts returned and reported Banks's column strung out and moving slowly down the Valley Turnpike with a Federal detachment picketing the Front Royal road to Strasburg.

On the morning of May 23, Jackson sent Ashby's cavalry east to block the Manassas Gap Railroad line running into Strasburg, and then brought his entire strength against Colonel John R. Kenly's 1,000-man Federal outpost near Front Royal. As the attack commenced, Confederate spy Belle Boyd arrived on a lathered horse, located Brigadier General Richard Taylor at the head of the Louisiana Brigade, and provided him with complete information on Banks's dispositions. Taylor, the son of President Zachary Taylor and brother-in-law to Jefferson Davis, did not wait for Jackson and launched an immediate assault. Jackson arrived and turned loose a second brigade in support of Taylor.

Kenly reported the attack to Banks but said he could hold the position without realizing that 17,000 Confederates were about to swallow up his command. During the afternoon, Kenly fell back to keep from being cut off, but too late. The Valley Army lost about fifty men, but the Federals lost 904 killed, wounded, or captured, including Kenly, who suffered a severe wound.

Having interposed the Valley Army between Banks and Washington, Jackson seized the initiative. Banks no longer had strong reserves at Winchester or Harpers Ferry to alleviate his predicament. A new race began. As Banks fled north with about 7,000 men, Jackson's foot cavalry pursued. Two days later, they clashed at Winchester.

The ripple effect of Banks's defeat at Front Royal produced consequences for McClellan's aggressive efforts on the peninsula. The Union war department rescinded orders for the Army of the Rappahannock at Fredericksburg to cooperate with McClellan and ordered Major General Irvin McDowell to prepare to move in Banks's defense.

THE THREE-DAY RACE

After the Front Royal fight, Jackson split his force to move more swiftly in an effort to beat Banks to Winchester, but he wanted more reconnaissance. At daybreak on May 24, Ashby moved the Seventh Virginia Cavalry to Strasburg. Another cavalry detachment with Hotchkiss as scout rode toward Middletown, a village five miles north of Strasburg and thirteen miles from Winchester. The Second and Sixth Virginia cavalries moved toward Newton, using the Front Royal–Winchester road. Ewell moved to Cedarville on the same road to await reconnaissance reports and marching orders. The remainder of the command moved to Cedarville and took position on the road leading west to Middletown.

Federal sharpshooters fired on Hotchkiss as he approached Middletown. Banks had deluded himself into believing the fight at Front Royal, despite heavy casualties, had been a raid by a small Confederate detachment and that Jackson's main force was around Harrisonburg. As a consequence, Banks did not begin the march to Winchester until 10:00 a.m. Hotchkiss returned to Jackson's command post and reported Banks's corps fanned out on the Valley Pike from Strasburg to Middletown. Leaving Ewell near Cedarville, Jackson brought infantry forward, chased off the sharpshooters annoying Hotchkiss, and climbed to a rise just east of Middletown. Scanning the pike as it stretched through Middletown, Jackson observed a seemingly endless line of infantry, cavalry, and wagon trains plodding slowly north.

Taylor's Eighth Louisiana moved double-quick into the Middletown area, fired on a passing cavalry

Shenandoah Valley: Strasburg to Winchester [Map #89]
Hotchkiss continued to build and refine the Shenandoah Valley map throughout the war. This section, from Strasburg to Winchester, shows a section of the Valley Turnpike stretching from Strasburg through Middletown, Newtown, and Kernstown to Winchester. Farther to the east is another road running out of Front Royal through Cedarville and Nineveh to Winchester. Hotchkiss mapped all the roads, rivers, towns, and terrain features on his Shenandoah Valley map. Jackson's Valley Army marched up and down this pike, from Winchester to Port Republic, three times during the Shenandoah Valley campaign. The map overlaps the segment on page 39 [Map #89-2].

company with deadly precision, and sent the Federals scrambling in every direction. Jackson took possession of the deserted wagons and sent cavalry and infantry down the pike in pursuit of Banks's fleeing army. At 3:00 p.m. Jackson sent a courier to Ewell with orders to take the Front Royal road from Cedarville and move to Winchester as quickly as possible. Jackson then put his entire force on the Valley Pike and sent them after the disorganized horde of blueclads retreating toward Winchester. The Valley Army might have overtaken Banks that day had Ashby's undisciplined cavalry and Taylor's Louisiana Tigers not abandoned the chase to loot captured wagons.

Although Jackson fretted at the delay, he wasted no time with recriminations and tried to press his brigades forward. As the blueclads retreated, Colonel George H. Gordon's Federal brigade burned wagons and attempted to delay Jackson's force on the pike. Had it not been for Gordon's stubborn resistance, Banks may never have reached Winchester with his force still organized. Confederates and Federals skirmished late into the night, firing at each other from the light cast by burning wagons. By the time Jackson's infantry reached the outskirts of Winchester on the morning of May 25, many of the men had marched more than twenty miles and slept no more than an hour since the previous day. Hotchkiss also felt fatigued, but he witnessed Jackson's determination to press the march until reaching the hills rising above the southern outskirts of Winchester.

Brigadier General Richard Taylor (1826–79), *son of former president Zachary Taylor, became an outstanding combat officer while serving under Jackson. Taylor eventually returned to Louisiana with his division and distinguished himself as a lieutenant general during the Red River campaign.*

While Jackson's forces remained hidden behind Massanutten Mountain, Hotchkiss took scouts to the summit to monitor Federal activity and reported Federal cavalry dashing up and down the Valley Pike in search of the Confederates.

THE BATTLE OF WINCHESTER: MAY 25, 1862

During the night of May 24–25, while Jackson pushed his exhausted troops relentlessly down the Valley Pike, Ewell's division moved on Winchester from the Front Royal road. At dawn, before advancing to take possession of the critical hills, Jackson sent an officer to locate Ewell's column, which was still moving into position. His plan of action was simple. With Arnold Elzey's brigade in reserve, the brigades of Winder, Campbell, and Taliaferro would hold Gordon's Federals in position, and when Ewell attacked the Federal left, Taylor would move around the enemy's right and assault Banks's flank.

Still unaware of the size of Jackson's force and having overlooked Ewell's division, Banks decided to take advantage of Winchester's fortified earthworks and do battle. He posted Gordon's brigade on the right and Colonel Dudley Donnelly's brigade on the left.

Although unaware of Banks's dispositions, Jackson gave Winder's Stonewall Brigade the task of leading the assault. With Campbell's brigade in support, Winder's grayclads splashed across Abraham's Creek and drove in blueclad pickets on the Union center. Neither brigade ran into determined resistance on the first ridge but afterward collided with Donnelly's brigade and encountered heavy Federal artillery fire from eight rifled guns perched on hills. Jackson now obtained a good view of Banks's defenses. He brought up three batteries and sent Taylor's Louisiana brigade along a concealed route to assault Banks's right flank. The artillery duel lasted thirty minutes with heavy losses before the Confederate batteries drove the enemy's guns from the hills.

Acting on his own, Ewell sent Brigadier General Isaac R. Trimble's brigade onto the field on the Federals' far left. At 8:00 a.m. Ewell opened with artillery and shattered the stone walls behind which the enemy had been checking the advance of Trimble's brigade. Trimble moved forward from behind several small hills to charge the enemy's left flank at the same time that Taylor struck Gordon's right flank. Although Gordon anticipated Taylor's attack and moved two Federal regiments to meet it, Banks overlooked the advance of Trimble on the left. Taylor and Taliaferro assaulted in force, Elzey added his brigade to Winder's frontal attack, and Trimble advanced on the Federals' left flank and entered Winchester unchecked. Within minutes, the entire Federal line collapsed. For the first time in the valley, the hideous rebel yell probably startled the blueclads more than the sunbeams dancing off 10,000 bayonets.

The Federal stampede happened so quickly that Jackson did not have time to organize a pursuit. He looked for Ashby but could not find him because several cavalry companies had recklessly charged, staggered back from withering volleys, and rode to the rear on wounded horses to re-form. Brigadier General George H. "Maryland" Steuart, commanding a brigade of cavalry, waited two hours before pursuing Banks because he did not receive orders from Ewell, to whom his command had been attached. However, he eventually responded to an order from Jackson and began to round up prisoners. Had Ashby been on hand, Jackson very likely would have destroyed Banks's army.

During the campaign that started in early May, Jackson's force suffered only 400 casualties. Banks lost control of the Shenandoah, 3,500 soldiers, 9,354 small arms, 500,000 rounds of ammunition, huge stores of sorely needed medical supplies, 103 head of cattle, 14,637 pounds of bacon, and tons of food and supplies that Major John Harman, Jackson's chief quartermaster, estimated to be worth a half million dollars.

The Taylor Hotel in Winchester became the headquarters for both armies, depending on which army occupied the town. Hotchkiss spent several nights at the hotel whenever the Confederates came to town.

Battle of Winchester
[Map #85-2]

Not all of Hotchkiss's maps went into his archival collection. He made several maps that were submitted to General Lee with Jackson's after-action battle reports. Although some sketches have survived from the battle of Winchester on May 25, 1862, this later, redrawn map accompanied Jackson's account for the Official Records.

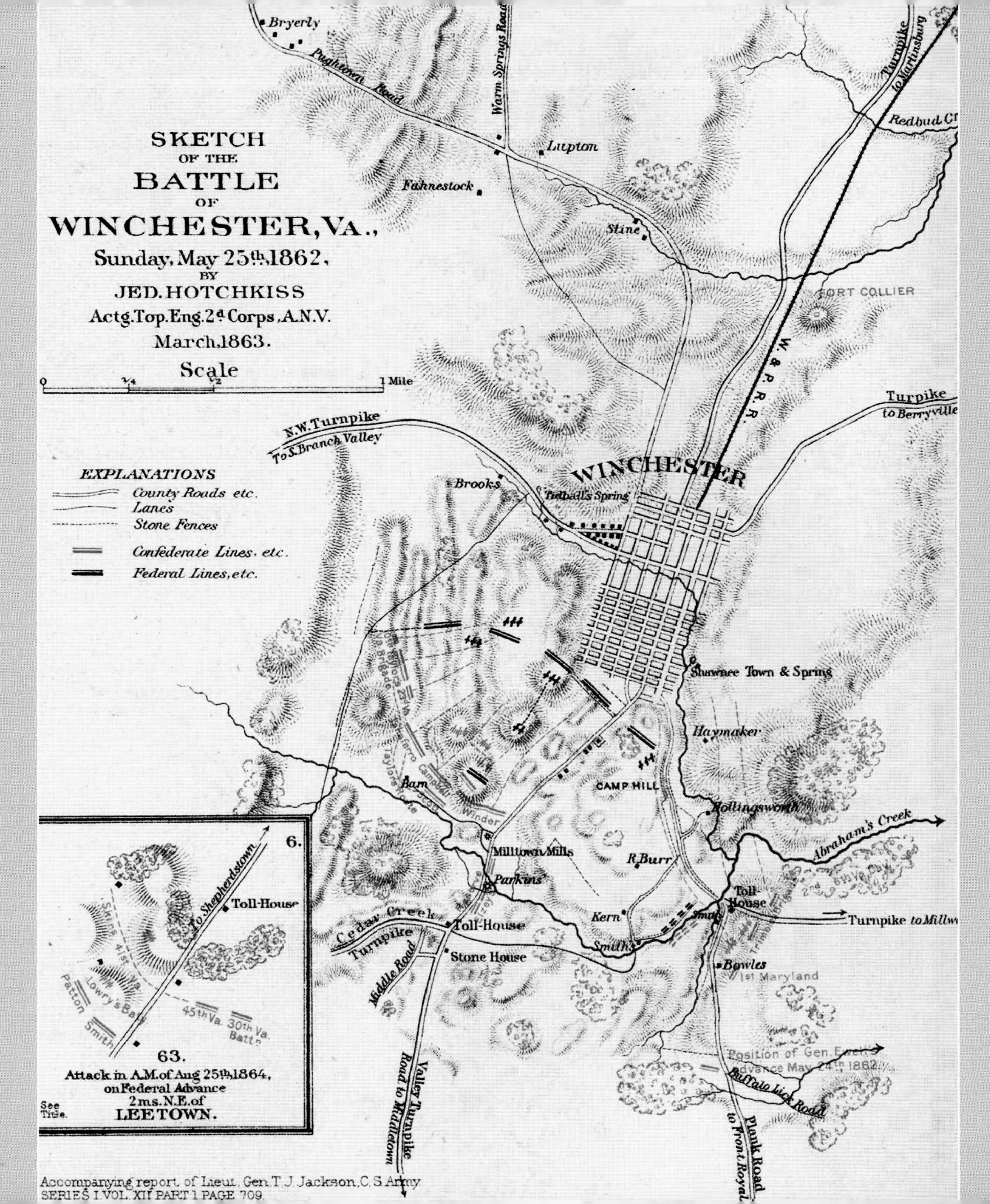

THE ESCAPE OF THE FOOT CAVALRY

The Confederate capture of Winchester on May 25 occurred at the same time that General McDowell's army was laying pontoons across the Rappahannock at Fredericksburg to join forces with General McClellan, whose Army of the Potomac on that day began shelling Richmond.

Having lost the opportunity to snare Banks, Jackson rested the army for two days at Winchester. On the second day, he sent Winder's Stonewall Brigade north in pursuit of Banks, mainly to drive the Federals into Maryland. Jackson had no intention of crossing the Potomac, but he was curious to see how Washington would react to the threat.

On May 27 he learned that General McDowell's movement at Fredericksburg to join McClellan had been canceled and up to 30,000 Federals, nearly twice the number Jackson commanded, were marching westward on the road to Front Royal. He also learned that another 15,000 blueclads from General Frémont's Mountain Department had been ordered by President Lincoln to march to Strasburg. Although the actual numbers detached by McDowell and Frémont were overstated, the Union strategy made sense. If Shields's division from McDowell's command and Frémont's force blocked Front Royal and Strasburg, Jackson would be trapped in the lower valley. A third Federal force was being hurriedly assembled at Harpers Ferry under Brigadier General Rufus Saxton Jr. with possible help from Banks at Williamsport, Maryland, and would close the pincers on Jackson's army and put the Confederates out of business. Should Shields require assistance, McDowell also moved Major General Edward O. C. Ord's division to the Manassas Gap railroad, where transportation would be available in an emergency. The success of the Federal strategy depended on which force moved the fastest.

Jackson recognized the threat and intended to slip through the portal before it closed. He sent the main army back through Winchester and on to Strasburg. On May 31 he recalled the Stonewall Brigade from Charlestown, where Winder's scouts at Halltown and on Loudoun Heights had been observing the buildup of Federal forces at Harpers Ferry. At 3:00 p.m. he explained his plans to Hotchkiss: "I want you to go to Charlestown and bring up the First Brigade. I will stay in Winchester until you get here if I can, but if I cannot, and the enemy gets here first, you must bring it [Winder's brigade] around through the mountains." Although many of the men in the Stonewall Brigade were familiar with the roads between Charlestown and Winchester, nobody had studied the roads in western Virginia more than Jackson's topographer.

Hotchkiss rode through the night and reached Winder at first light. The general pulled in the scouts and the pickets and put the brigade on the road. Rain fell in sheets and turned the highways to mud. Fearing that any delay might entrap the brigade, Hotchkiss pushed on and reached Winchester at dusk. Jackson had already moved to Strasburg but had left orders for Hotchkiss and Winder to keep moving. By then the Stonewall Brigade had marched thirty-five miles on mud-slick roads, and most of the men had consumed all their rations at breakfast. At 10:00 p.m. Winder realized his men could go no farther and called a halt at Newton, eight miles south of Winchester. Hotchkiss slumped on a pile of hay in a nearby barn and slept beside his horse.

At daylight on June 1, the brigade formed a column and marched the final twelve miles to Strasburg. Jackson ordered Winder to push on. Shields's cavalry had already reached Front Royal, and Ashby's bedraggled cavalry had formed the rear guard. By the time Shields's division reached the Valley Pike, Jackson's army had escaped without loss. The Confederates—and Winder's brigade in particular—had marched twice as far as the Federals in half the time.

Frémont's army arrived at Strasburg much later. His column encountered difficulties moving through western Virginia because, two weeks earlier, Hotchkiss, working with Sterrett's cavalry, had blocked the mountain roads in passes that provided the most direct routes to the valley.

Shenandoah Valley: Strasburg to Mt. Jackson [Map #89-2]
This section of the Hotchkiss Shenandoah Valley map connects to the preceding section on page 34 [Map #89-1]. It includes the area where Hotchkiss first began the valley mapping project for General Jackson. Strasburg, which connects by road with Front Royal to the east, is where Jackson outwitted General Banks on May 23 and General Shields on June 1.

STRASBURG
MAURERTOWN
WOODSTOCK
EDENBURG
JACKSONVILLE
MILFORD
SPRINGFIELD
HAWKINSTOWN
MT JACKSON
SHORT
STONY CREEK

CHARLES SIDNEY WINDER (1829–62)

Born in Talbot City, Maryland, Winder was one of the few West Pointers (class of 1850) in Jackson's Valley Army. He entered the Confederate service as a major in charge of artillery and participated in the bombardment of Fort Sumter in April 1861. Promoted to colonel of the Sixth South Carolina Infantry Regiment, Winder's association with the Stonewall Brigade happened by chance when Jackson arrested Richard B. Garnett for withdrawing the brigade at Kernstown. Jackson needed a brigadier general to command the brigade, and the war department sent Winder.

The decision to not put a Virginian in charge of the Stonewall Brigade enraged its members and, in particular, its officers. While regimental commanders accepted Jackson's style of West Point discipline, they thought it demeaning to accept the same type of treatment from a Marylander. Officers greeted Winder coolly, and enlisted men hissed when he rode by on his horse. Some in the ranks even threatened to kill him during the next big battle. Winder shrugged off the carping and went about assiduously rebuilding the brigade.

Winder led the Stonewall Brigade with distinction in all the battles of the Shenandoah Valley campaign. Although his tactics and determination did not win the brigade's affection, it won their respect. He also refused to allow Jackson to criticize the brigade without good reason, and during one confrontation drew a cheer from the men. During the campaigns of 1862, he continued to grow in stature as a fighter, and when he was mangled by an artillery shell and killed during the battle of Cedar Mountain on August 9, 1862, his death was lamented by Jackson and General Robert E. Lee—but not by the Stonewall Brigade.

DESIGNING A NEW STRATEGY

President Lincoln's war department had been quite specific when issuing orders to General Shields and General Frémont. They were to pursue the Valley Army and dispose of it before Jackson caused more trouble. "Old Jack" expected to be relentlessly pursued by a powerful force twice the size of his own army and continued to press his brigades up the valley. Jackson also understood the fallacy of the Union command. Frémont and Shields both operated independently with no one in overall command, making it almost impossible to coordinate a common strategy.

On June 4 Jackson slowed the pace of the Confederate withdrawal as the long column approached Harrisonburg. The region suited him because Hotchkiss lived nearby and was intimately familiar with the terrain. Twice during the day, Jackson consulted Hotchkiss on the topography around Port Republic.

Early the following morning, Jackson dispatched Hotchkiss to Peaked Mountain with a signalman and instructed him to reconnoiter Shields's division in the Luray Valley and any movements of Frémont's army on the west side of the mountain. Hotchkiss spent most of the day strolling the summit. He observed the head of Shields's camp near Conrad's Store, with the balance of the division encamped near Luray. He saw no evidence of Frémont's army, which Ashby recently reported miles away on the Valley Pike.

Hotchkiss looked for Jackson in Harrisonburg and learned that he had gone to Port Republic. He finally located the general and informed him of Shields's position on the east side of the Massanuttens. Jackson already knew that Frémont lagged behind on the opposite side of the mountains and had not joined forces with Shields. Jackson considered this a rare opportunity and made the bold decision to hurl his entire force at Shields's 10,000-man division and then pick the right opportunity to turn the same tactic on Frémont's 11,000-man army, thereby using his superior force to defeat one Federal command before defeating the other. With this in mind, Jackson sent Hotchkiss and a signalman back to Peaked Mountain. On the morning of June 6, Hotchkiss stayed below to receive the signalman's messages from the summit, which he then sent by courier to Jackson.

Not everything went as Jackson planned. His infantry were thoroughly exhausted and the cavalry had not enjoyed a moment's rest since reaching Harrisonburg. During the day, numerous cavalry clashes occurred on the Port Republic road, and there were more signs of enemy infantry activity in the area. Late that evening and two miles south of Harrisonburg, Ashby took a regiment of Confederate infantry into the woods to cut off an enemy cavalry probe. As dark approached, the infantry began to shoot at anything that moved. Ashby told them to hold their fire while he went forward to reconnoiter. One grayclad did not get the message and killed Ashby by mistake. Despite Jackson's disagreements with Ashby over professional military behavior, he considered his cavalry commander indispensable. He referred to Ashby's death as "a loss irreparable," and it came at a most inopportune time for Jackson. According to Hotchkiss, "Gen. Jackson walked the floor to his room . . . in deep sorrow, greatly moved by the sad news."

Hotchkiss spent the morning of June 7 reviewing his maps of the Port Republic area with the general. When the signalman on Peaked Mountain indicated the approach of Frémont's infantry, Jackson offered battle near Cross Keys, trying to draw the Federals into a fight with Ewell, but the enemy withdrew. Frémont had pressed forward with just his cavalry and a small detachment of infantry and showed no willingness to engage in battle. As the sun set that night, Jackson did not know whether he would be fighting in the morning against Frémont or Shields. According to Hotchkiss's last report, Shields was still at Conrad's Store, a day's march away. Jackson hoped the two Union columns would remain separated, but time was running out.

During the battle at Cross Keys, fatigued Confederates sought cover and tried to obtain a small amount of rest while reloading their rifles for the assault.

Shenandoah Valley: Harrisonburg to Port Republic [Map #89-4]

Another key section of the Shenandoah Valley map concerned the area between Harrisonburg and Port Republic, where the Valley Army defeated General Frémont's forces on June 8 and General Shields's forces on June 9. Although Hotchkiss had produced sketches of the area prior to the two battles, he did not complete this section of the map until afterward.

THE BATTLEFIELD

Hotchkiss spent hours discussing the Port Republic area with Jackson because it was uniquely adapted to fighting two hostile armies with one. The South Fork of the Shenandoah River began at the confluence of the North and South rivers, which ran along both sides of Port Republic before merging.

A bridge crossed the North River just above the junction with the South Fork, over which the Harrisonburg road entered Port Republic. The South River, which cut off Port Republic from the Luray Valley, had no bridge but could be crossed at two hazardous fords.

North of Port Republic and on the left bank of the Shenandoah, a line of high bluffs covered with scattered timber completely commanded the tract of open farmland lying between the river and the Blue Ridge, and through this low ground ran the road on which Shields had been marching.

Four miles northwest of Port Republic, near the village of Cross Keys, the road from Harrisonburg crossed Mill Creek, a strong position for defense.

Jackson clearly understood from Hotchkiss's maps that by burning the bridge at Port Republic, he could easily escape from Frémont and assail Shields in the Luray Valley with a superior force. Jackson did not like this option because Frémont could still bring his artillery to the banks of the South Fork to support Shields. Jackson's other option was to use part of Ewell's division to block Frémont's advance by defending a ridge that ran on both sides of the Harrisonburg road.

Jackson had one more option. He could defeat Frémont first, but assuming that Shields had the more powerful force, he preferred to eliminate Shields before dealing with Frémont. If Shields attacked while the Confederates were engaged with Frémont, the rear of Jackson's position would be exposed and his communications threatened. Any assault on Frémont first would make the Valley Army instantly vulnerable to an attack by Shields. After contemplating his options, Jackson placed most of Ewell's division on the ridge over the Harrisonburg road to block Frémont and kept the Stonewall Brigade encamped near Port Republic to deal with Shields. Not everything worked as Jackson planned.

THE OPENING ACTION

At 9:00 a.m. on the morning of June 8, a reconnaissance in force by Brigadier General Samuel S. Carroll's infantry brigade from Shields's division entered Port Republic and brushed away Confederate pickets. Jackson heard sounds of firing and, without bothering to saddle his horse, walked toward it. He had not expected to be attacked by Shields so soon and grew concerned because the rear of his army was defenseless. He also worried that Frémont and Shields had finally come together and had agreed to conduct a joint operation.

As soon as his mount arrived, Jackson rode into town with five members of his staff, one of whom was Lieutenant Colonel Stapleton Crutchfield. Jackson's party started toward the crucial lower ford over the South River and inadvertently became involved in the action. Federal cavalry captured Crutchfield and another staff officer, and Jackson barely escaped to the other end of town.

What appeared to be the beginning of a major engagement gradually withered to a heavy skirmish. Jackson took command at the far end of town, brought Captain S. C. Moore's company across from the upper ford of the South River, stationed a pair of Captain James Carrington's howitzers at the end of town, and drove off the enemy. General Carroll's four regiments of blueclad infantry made the mistake of forming behind Federal batteries. When Carrington's guns opened fire, some of the shells overshot the enemy's batteries and ripped through Carroll's infantry. When the blueclads observed the Federal cavalry in full retreat, Carroll's front collapsed and the entire force withdrew northward into the Luray Valley. Jackson correctly surmised that Shields had sent a reconnaissance and the main body of the Federal division had not left camp. He put two brigades in position to guard the South Fork bridge and turned his attention to reports of fresh fighting at Cross Keys.

Hotchkiss heard the firing, but he had collapsed from a severe headache brought on by fatigue and could not get on his horse without feeling faint.

What concerned Jackson was Frémont's seemingly simultaneous attack at Cross Keys, which was located off a side road about seven miles northwest of Port Republic near Keezletown. Ewell had moved toward Cross Keys with only three brigades—about 6,000 infantry and 500 cavalry under Elzey, Trimble, and Taylor. Frémont's force consisted of 10,000 infantry, twelve batteries, and about 2,000 cavalry. Jackson expected Ewell to hold his ground if attacked, and the sounds of skirmishing at 8:30 a.m. became troubling. With Shields threatening, Jackson felt compelled to keep his main force at Port Republic.

Ewell had spread a number of Confederate regiments along the crest of a narrow ridge overlooking an open meadow with a small creek. The ground occupied by the Confederates was only partially cleared with scattered openings in the woods for placing artillery. Frémont's forces occupied the lower ridges at Cross Keys, where patches of heavy woods concealed Federal movements. At 10:00 a.m. the engagement erupted in earnest when Frémont opened with artillery.

Frémont deliberated because of misgivings. He agonized over taking the offensive against Jackson's army, and he had no maps of the area. Nor had he communicated with Shields, and he feared he might be confronting the Army of Northern Virginia and not just Jackson. Frémont worried more about losing a battle than winning one. As a consequence, he chose the poorest tactics possible and resorted to half-measures. Had he hotly pressed the attack in an effort to paralyze Ewell, the outcome might have been more valuable than a victory.

With twenty-four regiments on the field, Frémont sent forward only five from Blenker's German division. Blenker struck Trimble's brigade on the Confederate right, possibly the worst place on the battlefield to begin an attack. Trimble's brigade occupied a flat-topped ridge covered by enormous oaks that looked down on the meadow through which Blenker advanced. Confederate skirmishers drew the enemy forward, luring them toward a thicket, and when Blenker's regiments came within sixty yards, grayclads opened with a sheet of fire that drove the Germans careering back across the meadow.

Trimble's brigade gave a token chase and fell back to their original position while Blenker ordered up his reserves. When Frémont showed no inclination to resume the attack, Trimble used heavy cover in a ravine to work around Blenker's left flank. Reinforced by six regiments from Ewell, Trimble struck Blenker's left wing and threw eleven Federal regiments back in disorder.

Frémont made a feeble attempt to breach the Confederate center, but his regiments were routed by one of the few bayonet charges of the war. On Frémont's right wing, Milroy's and Schenck's brigades, which Jackson had defeated at McDowell, advanced on their own initiative. Just as the Federals began to make progress against Taylor's brigade, Frémont recalled Milroy and Schenck to help shore up the center against Trimble's assault.

Jackson came on the field but let Ewell fight the battle unimpeded. He understood Frémont and merely suggested to Ewell, "Let the Federals get very close before your infantry fire; they won't stand long." As darkness approached, Frémont withdrew and Ewell occupied the position previously held by the Federals. During the night, most of Ewell's force slipped back to Port Republic, leaving only one regiment at Cross Keys to keep an eye on Frémont.

Battle of Cross Keys [Map #95]
Hotchkiss copied a sketch from his archive and gave it to Ewell. The general might have found Cross Keys without any help from Hotchkiss, but the area was not spelled out on any map, and Early was completely unfamiliar with the roads turning off the main highway. The sketch also proved helpful in Ewell's selection of a superior defensive position.

One of the problems in communications between generals Frémont and Shields was caused by Confederates operating far in the Federal rear. They cut Federal telegraph communications almost as soon as the lines were strung.

BATTLE OF CROSS KEYS CASUALTIES

	KILLED	WOUNDED	MISSING	TOTAL
CONFEDERATE	41	232	15	288
FEDERAL	114	443	127	684

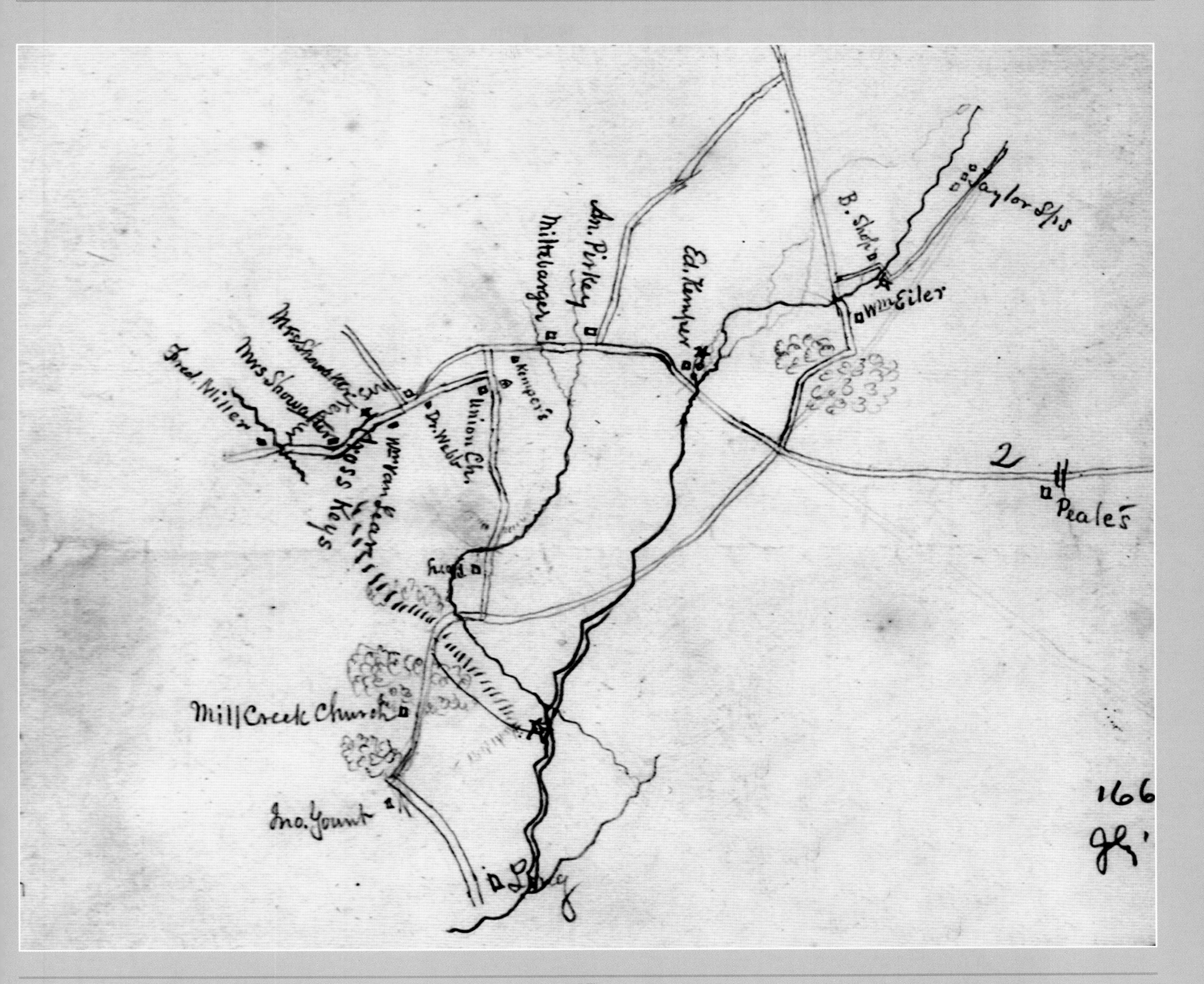
An. Pirkey
Miltebarger
Ed. Kemper
B. Shop
Taylor Shps
Wm Eiler
Peale's
Union Ch.
Dr. Webb
Kemper's
Mrs Showalter
Fred. Miller
Cross Keys
Mill Creek Church
Jno. Yount
166

BETWEEN DARKNESS AND DAWN

Hotchkiss recovered from his headache late in the day and, feeling better, rode to Cross Keys to watch the battle. He returned with Ewell to Port Republic during the night. When the troops crossed over the North River, Jackson gave orders to burn the bridge after the rear guard crossed in the morning. Destroying the bridge had been part of Jackson's original plan to prevent Frémont from joining Shields. Meanwhile, a footbridge of wagons had been thrown over the South River so that Ewell's troops could cross and take position near the home of General Lewis, where Jackson planned to concentrate his forces for a morning engagement with Shields's division.

During the evening, Shields received a belated message written by Frémont the previous day suggesting a joint attack. Aware that a battle had commenced at Cross Keys and lasted throughout the day, Shields believed the Confederates would have been severely weakened and decided, as Jackson correctly anticipated, to attack the Confederates in the morning.

At 2:00 a.m. on June 9, Jackson called his brigade commanders and Ewell to headquarters to explain his orders. Hotchkiss observed that just before Jackson expected to go into battle, he always spoke to his commanders to ensure that his instructions were thoroughly understood, and he never did so until he was certain that those instructions would not fall into the hands of the enemy. He gave Trimble and Colonel John M. Patton the task of holding off Frémont's entire army while the main Confederate force tackled Shields.

Jackson did not wait for developments. At 5:00 a.m. Winder advanced with four regiments from the Stonewall Brigade through a storm of picket firing, which throughout the night had grown in intensity. Dressed in full uniform, Jackson led the brigade into the Luray Valley. Taylor's Louisianans followed half a mile behind. To some of the men, it appeared that Jackson intended to fight Shields's 9,000-man division with sixteen guns and less than 4,000 infantry.

After driving in Federal pickets, Jackson obtained a good look at Shields's dispositions. Hotchkiss arrived at Jackson's side and was asked where he had been. Hotchkiss began explaining his illness when heavy artillery fire began taking lives in the Stonewall Brigade. Jackson nodded to Hotchkiss and said, "Take General Taylor around and take those batteries." Part of the Federal artillery was on the plain near the Lewis house, with another section near the river. A third battery fired from a shoulder of a ravine near Fletcher's farm and swept the plain. Shields had posted his artillery well. The men who served the guns were of a different caliber than Blenker's Germans. The Federal officers were veterans and also made of stauncher stuff. They had whipped Jackson at Kernstown and were prepared to do it again. Unlike Frémont's officers, none of them dreaded the name Stonewall Jackson.

Jackson had seen the battlefield and Shields's forces, and now he felt compelled to get back to Port Republic to see what Frémont was doing. He told Winder to attack, go into the woods on the right, and outflank the battery beyond the stream near the Lewis house. Winder collided with a strong force of Federals in the woods, fell back, and became isolated. Taylor was still moving forward when Shields launched a counterattack and instantly checked the Confederate advance. Jackson stopped before reaching Port Republic and ordered up Ewell's division. Ewell could not get his men across South River quickly because the footbridge constructed of planks and laid across South River on the running gear of wagons began to break from the fastenings. General Taliaferro finally got his regiments across, moved up the western edge of the woods, poured two heavy volleys into the enemy's flank, and silenced some of the batteries that had pinned down Winder and Taylor.

Jackson returned to the field to rally the Stonewall Brigade. At the very moment it appeared the Confederates might be whipped piecemeal, Taylor brought up his reserves. With a piercing rebel yell, Colonel Kelley of the Eighth Louisiana led his companies across the ravine, swarmed up the bank occupied by Federal artillery, and, after two charges, carried the battery.

BATTLE OF PORT REPUBLIC CASUALTIES

	Killed	Wounded	Missing	Total
Confederate	89	951	45	1,085
Federal	67	393	558	1,018

Battle of Port Republic [Map #96]

Hotchkiss developed a topographical map of the Port Republic battlefield showing the terrain features rather than the placement of Confederate and Federal forces. By referring to the map while reading the narrative, the battle can be followed.

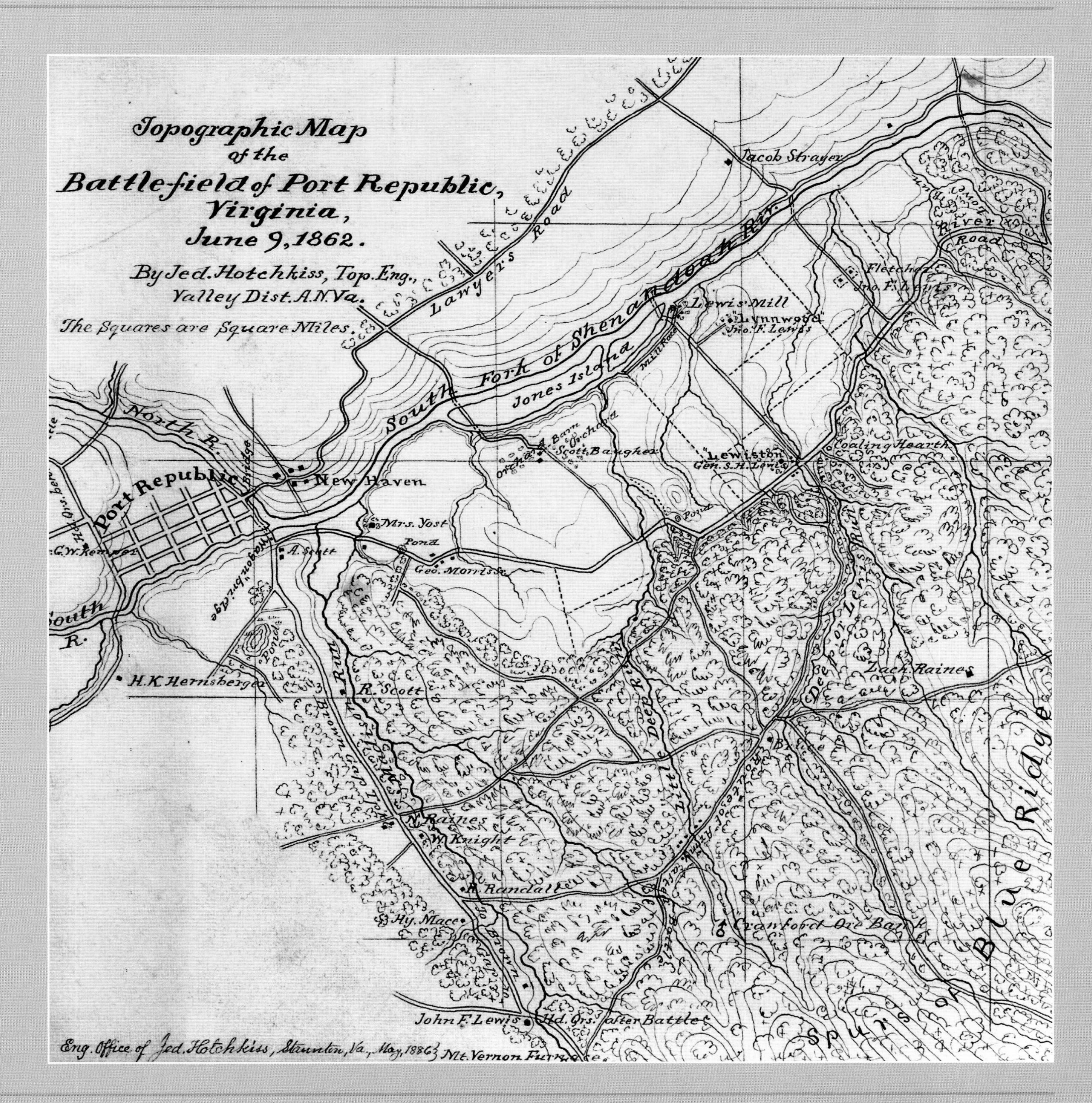

By then, the rest of Ewell's division had reached the field, cutting through the enemy and driving off the blueclads that had surrounded the Stonewall Brigade. Taylor turned the captured guns on the enemy, and in less that fifteen minutes drove General Tyler's Federal brigade from the field. Jackson insisted on pursuing the enemy, and chased them nine miles down the South Fork of the Shenandoah. The Confederate cavalry drove the blueclads to Federal headquarters at Conrad's Store, and there the chase ended when Shields formed two brigades and beat off Jackson's cavalry.

During the battle, Trimble's rear guard burned the bridge over the North River and rejoined Ewell. With the North River bridge burned, Frémont made no effort to cross the river and moved his artillery to the north bank of the South Fork of the Shenandoah. He made no further attempt to support Shields aside from firing at Confederate ambulances moving across the plain.

Around noon, Jackson recalled his entire force. He had another duty to perform, but not in the Shenandoah Valley.

THE AFTERMATH

Jackson remained in the vicinity of Port Republic no longer than it took the men to collect the wounded, bury the dead, and load the wagons. He ordered Hotchkiss to lead the army out of the valley on a route not likely to be observed by the enemy. Hotchkiss used byroads passing through woods to Mount Vernon (Miller's Furnace), then upward in a steady incline leading to Brown's Gap in the Blue Ridge Mountains. The army settled into Brown's Cove at the gap and rested for several days. The movement put Jackson in direct communication with General Lee's Army of Northern Virginia at Richmond.

Hotchkiss began a new project for Jackson, making detailed sketches of the surrounding area. He stayed with the rear guard at Mount Meridian two miles south of Port Republic to finish mapping the area and producing a battle map for Jackson's report. While there, scouts fanned out and reported that Frémont had withdrawn to Strasburg and Shields to Luray.

By then, the impact of Jackson's operations in the Shenandoah Valley had jolted Washington. The unexpected counterstroke, delivered by two Confederate divisions thought to be in full retreat, not only stunned Frémont and Shields but also shocked Lincoln and Secretary of War Edwin M. Stanton, who headed the Union war department. On June 6, the day Jackson halted at Port Republic, Brigadier General George A. McCall's division had been detached from McDowell's Army of the Rappahannock and transferred to McClellan on the peninsula. Two days later, General McDowell received new orders instructing him to move the balance of his army and form a juncture with McClellan. Stanton also cut orders for Frémont to occupy Harrisonburg and for Shields to occupy Fredericksburg. Before Stanton's dispatches reached their intended recipients, Jackson had defeated both Frémont and Shields and completely upset Federal plans. Stanton revoked his orders and went back to the drawing board.

Though far removed from the bloody battles about to commence on the peninsula, Jackson's Shenandoah Valley campaign substantially weakened McClellan by depriving him of reinforcements, which alternatively strengthened Lee. The speed of movement and the concentration of force so important to military success could not have been achieved without the invaluable assistance of Hotchkiss, who laid the entire topography of the valley before the general and showed him the way.

Though barely two months had passed since Hotchkiss joined Jackson, the general now had a topographer on his personal staff whom he considered to be indispensable.

After the Battle of Port Republic [Map #85-5]
Hotchkiss eventually created an after-action map of the battle of Port Republic that also shows the previous day's engagement at Cross Keys. A version of the map originally appeared in the Official Records*, series 1, volume XII, part 1, page 716, before being transferred to the atlas.*

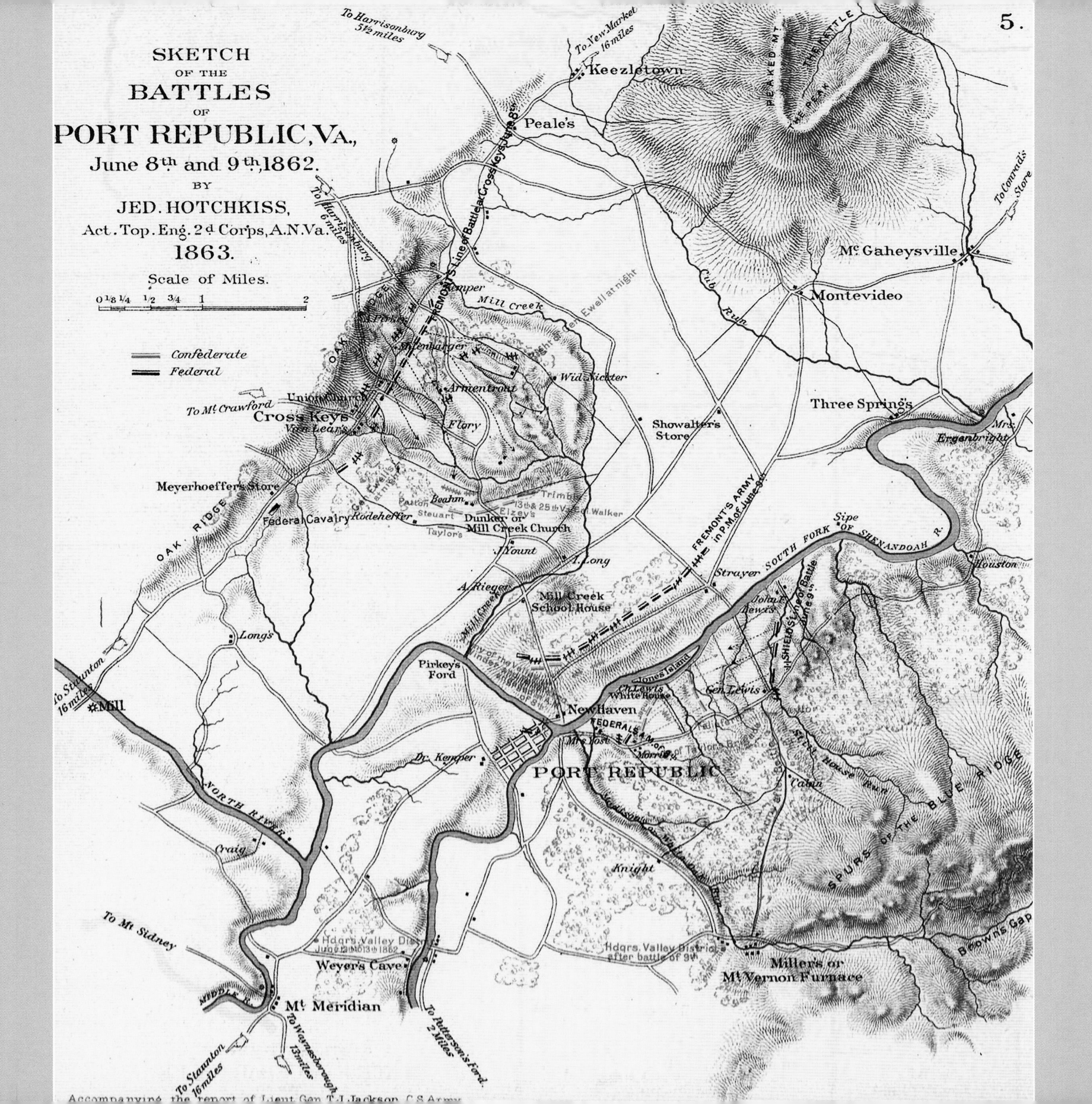

Accompanying the report of Lieut. Gen. T. J. Jackson C.S.Army

CHAPTER THREE: FROM RICHMOND TO ANTIETAM

After the battle of Port Republic on June 9, 1862, Jackson rested his men for nine days, but not his staff. Hotchkiss worked each day on maps of the area, sketches of the battlefield, routes through the gaps in the Blue Ridge Mountains, and routes to Lexington in the upper valley. A sense developed among the staff at headquarters that their secretive general planned to move from the valley. Reinforcements arrived on June 16, consisting of Brigadier General William H. C. Whiting's division and Brigadier General Alexander R. Lawton's brigade. Hotchkiss suspected that Whiting and Lawton had not come to fight in the valley.

Even more revealing were the number of times Jackson summoned Hotchkiss to bring maps. The meetings were usually private because of the general's practice to never expose his plans until he decided to act. As a result of the variety of maps Jackson requested, Hotchkiss suspected the general was more interested in specific maps but asked for others to conceal his intentions. The practice worked well until June 17, when Jackson asked Hotchkiss for maps of the Richmond area, where heavy fighting threatened the Confederate capital. Hotchkiss spent an hour with Jackson, enough time to conclude that the army would be marching. Perhaps because Jackson did not want his discussion with the topographer mentioned outside headquarters, he instructed Hotchkiss to pack his belongings and go back to Mount Meridian. After midnight on June 18, the army broke camp and marched toward Waynesboro. Thinking that he had been overlooked, Hotchkiss caught up with Jackson near Mechum's River Station with intentions of following the army wherever it went. The general promptly sent him back to Mount Meridian, saying that the topographer's time would be more valuably spent mapping the valley.

Hotchkiss appreciated the opportunity to move his tent closer to home, where he could visit his family on weekends, but he also worried about Jackson maneuvering on unfamiliar terrain. Jackson should have known better, as his effectiveness depended on having Hotchkiss at headquarters.

JACKSON AT RICHMOND

On May 31, 1862, after stalling for a month on the peninsula, McClellan's massive Army of the Potomac finally backed General Joseph Johnston's Confederates into the suburbs of Richmond. At the time, Jackson's Valley Army was still retreating toward Port Republic, Shields and Frémont were still in pursuit, and General McDowell's 40,000-man Army of the Rappahannock was still interposed between the Shenandoah Valley and Washington. On that day, Johnston's well-conceived Confederate plan for the battle of Fair Oaks and Seven Pines failed because of faulty execution resulting from poor planning by his staff. Although the Confederates pushed the Federals back a mile, Johnston suffered a severe wound. Jefferson Davis reacted by replacing Johnston with Robert E. Lee.

McClellan pressed forward cautiously, giving Lee time to strengthen Richmond's defenses. On June 12–16 Major General Jeb Stuart's Confederate cavalry rode completely around the Army of the Potomac and reported McClellan's right wing unsecured. Lee devised a plan to flank the enemy and ordered Jackson to march with his corps to Richmond. On June 24 Lee placed only 25,000 troops on the outskirts of Richmond to defend the capital against 75,000 Federals. He moved the other 47,000 troops into position to strike McClellan's vulnerable right flank on June 26. He also planned on that day for Jackson's Valley Army to arrive and spring a surprise assault on the Federal rear. Jackson arrived late, lost his bearings, and fought without his characteristic aggressiveness. During the following day's battle, he arrived late again. In all, Jackson failed four times to get into position on time and perform his designated tasks on the battlefield. Without Hotchkiss, the twists and turns of the Chickahominy River, the swamps and marshy forests, the bridges, and the chaotic network of meandering roads all combined to baffle the general. He simply became confused maneuvering through unfamiliar terrain without his guide because the existing maps, despite being on the doorstep of the commonwealth's capital, were all obsolete.

Jackson's presence, however, concerned McClellan. When the Confederates broke through the Federal position on June 27 and continued with assaults throughout the Seven Days campaign, McClellan believed his army was outnumbered and retreated to Harrison's Landing on the James River. The Confederates did not destroy the Army of the Potomac as Lee had hoped, but McClellan's campaign came to a quick and disastrous conclusion that isolated more than 90,000 Federal troops for nearly two months.

After detraining from the Shenandoah Valley, Jackson's foot cavalry begin the march to Mechanicsville during the Federal assault on Richmond in the Peninsula campaign. The typical Confederate was often shoeless and tattered from campaigning, but the infantryman always kept his musket clean and in good operating condition.

HOTCHKISS AND JACKSON REUNITE

On July 15 Hotchkiss received orders to gather his map collection and meet Jackson at Gordonsville, a small town about sixty miles northwest of Richmond. Hotchkiss knew the fighting had ended on the peninsula without knowing all the particulars. He expected to hear reports of a new campaign, and when no trains arrived at Staunton from Richmond, he suspected cars were being detained to move troops. Hotchkiss reached Gordonsville on July 18 and found Jackson's brigades camped in the area, but the general had not arrived.

Much had changed since the Peninsula campaign. Lee had reorganized Confederate forces into the Army of Northern Virginia with Lieutenant General James Longstreet in command of the First Corps and Jackson in command of the Second Corps. After the failure of Banks, Frémont, and McDowell to destroy Jackson in the Shenandoah Valley, Lincoln's war department unified the three commands and formed the Army of Virginia with 50,000 Federal troops under the command of Major General John Pope. When Frémont refused to serve under Pope, the war department replaced him with Franz Sigel.

Lincoln gave Pope three objectives. He expected the general to protect Washington and control the Shenandoah Valley. He also wanted Pope to move toward Charlottesville and draw Lee's strength away from Richmond. Pope took command of the Army of Virginia at the beginning of the Seven Days campaign, and after McClellan retired to the James River, Pope's key objective became the protection of Washington. On July 14 Pope began to slowly move the army toward Gordonsville.

Jackson arrived at Gordonsville on July 19 and issued no orders. Hotchkiss did not need instructions. He set up his tent and began "preparing maps for future use if not immediate." He observed that Jackson "looks the worse for his Chickahominy trip, and so do the troops." Hotchkiss and his staff, consisting of S. Howell Brown, Thomas O. Kinney, and Thomas Snead, pitched their tents on hillsides and began mapping the area to the north. A few days later, Major General Ambrose P. Hill's division joined Jackson at Gordonsville. Over the next three weeks, Lee consolidated the Army of Northern Virginia and rested the troops while the cavalry watched the movements of the enemy. Hotchkiss explored the area, riding as far north as Orange Court House, and used his staff to assemble sketches that eventually became Jackson's marching routes and battle maps.

ARMIES ON THE MOVE

On August 3, 1862, General Lee learned from scouts that Major General Ambrose Burnside's 14,000-man corps was moving up the Chesapeake on transports and disembarking at Aquia Landing near Fredericksburg to reinforce General Pope. Lee did not want Pope reinforced and gave Jackson a mandate to strike Pope before Burnside arrived. Jackson summoned Hotchkiss to discuss routes from Orange Court House to the Rapidan River that were least likely to attract the enemy's attention. Although the general had not consulted his division commanders, he planned to leave Gordonsville and march his Second Corps to Orange Court House on August 7, cross the Rapidan the following day, and attack Pope near Culpepper.

Once again Jackson frustrated all his subordinates by not sharing his plans until the last moment. Ewell had experienced Jackson's annoying penchant for secrecy during the Shenandoah Valley campaign. When a staff member asked where the army was headed, Ewell angrily replied, "I do not know whether we march north, east, south, or west, or whether we march at all. General Jackson has simply ordered me to have the division ready to move at dawn."

On the morning of August 7, Jackson's corps moved from Gordonsville to Orange Court House using back roads designated by Hotchkiss. That night, Jackson issued orders for the next day's march. Ewell's division would lead off on the main road to Culpepper, followed by A. P. Hill's and Winder's divisions. During the night, Jackson changed his mind and told Ewell to march by a different route, but he failed to inform Hill and Winder. In the morning, Hill waited for Ewell, who had gone by a different route, and became furious when Winder's division appeared on the road instead. Hill fumed and waited. Jackson appeared on the field and blamed Hill for the delay. Hill blamed Jackson. Jackson halted the march, dispatched orders for Ewell to stop, and bivouacked another night at Orange Court House.

The delay could have been costly had Pope not been slow in positioning his troops between Culpepper and Warrenton to block the main routes to Washington. Only Banks's 11,000-man corps, the vanguard of Pope's army, was on the road and eight miles south of Culpepper when Brigadier General Beverly H. Robertson's Confederate cavalry, screening Jackson's movement, fell upon Federal scouts. On the route behind Robertson came Ewell's division, followed by Winder and Hill, a total of 24,000 men and 1,200 wagons occupying seven miles of road.

Battle of Cedar Mountain [Map #85-4]

The battle of Cedar Mountain went by several names. Hotchkiss called it Cedar Run; others called it Slaughter Mountain, which was the proper name for the heights where Ewell placed his artillery. On the day following the battle, Hotchkiss sketched more of the terrain and began collecting information for a map later submitted with General Jackson's battle report. Hotchkiss suggested the map be named Cedar Run because the action actually occurred near the stream. The general agreed, but historians have routinely referred to the engagement as the battle of Cedar Mountain.

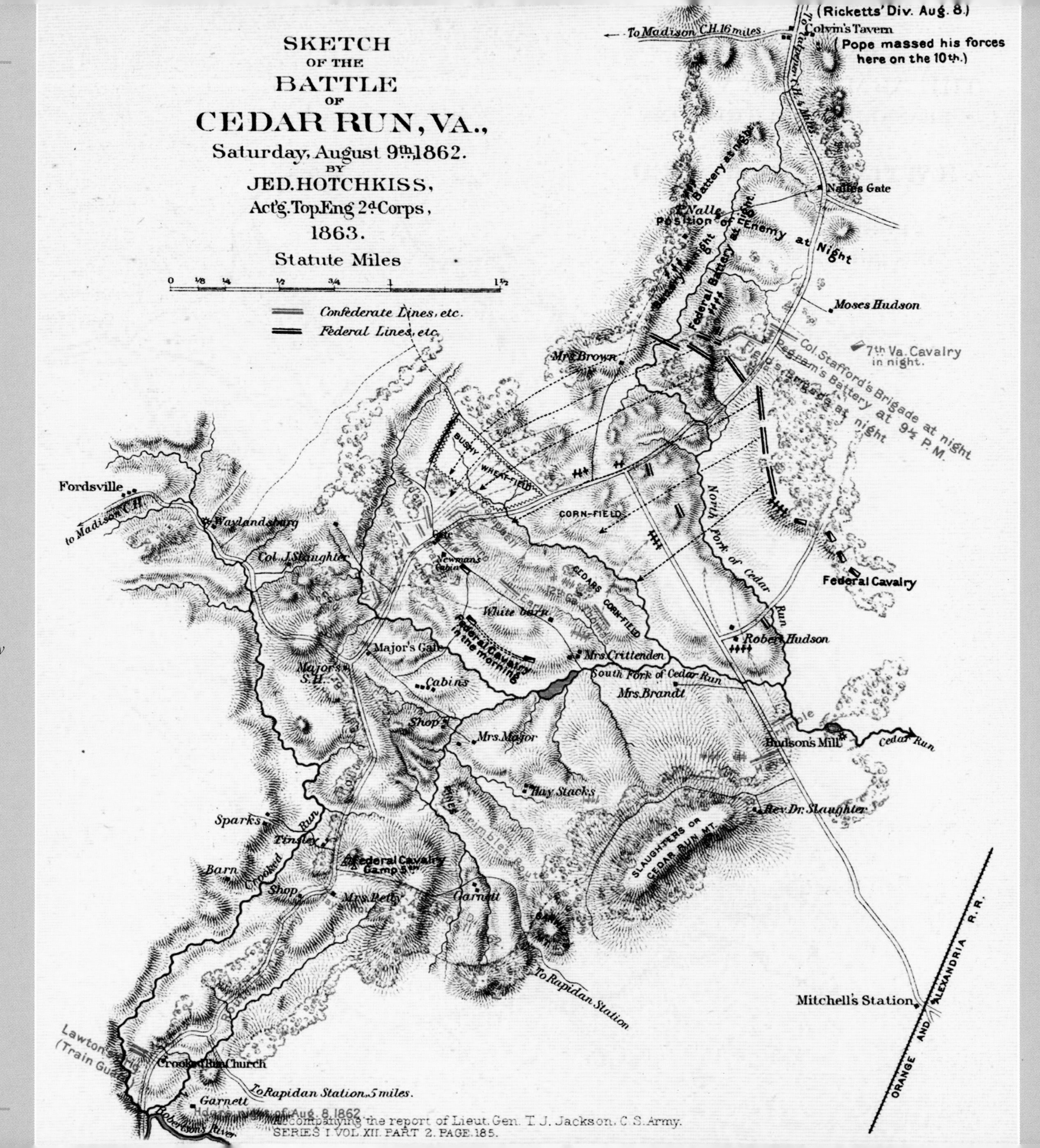

At noon, when Jackson learned that the enemy occupied the road to Culpepper, his men were already collapsing from sunstroke in temperatures approaching a hundred degrees. Winder rode in an ambulance. He had been ill and also suffered from the heat. Jackson rode ahead as Federal artillery opened on Ewell's lead brigade, commanded by forty-four-year-old Brigadier General Jubal A. "Old Jube" Early. A pugnacious fighter, Early had reached the western slope of Cedar Mountain, a low, mile-long ridge, when Federal shells began to descend on his column. Jackson ordered Ewell to secure the heights, place the artillery along the northern base of the mountain, and position Early's brigade to face the Federals advancing on the road. Ewell kept Trimble and Colonel Henry Forno's brigades in reserve and placed them with the artillery.

Brigadier General Isaac Ridgeway Trimble (1822–88) *commanded a brigade in General Ewell's division in the Shenandoah Valley campaign. Hotchkiss worked closely with Trimble during later battles at Cedar Mountain, Second Manassas, and Gettysburg.*

Winder emerged from his ambulance against the doctor's orders to position his division on Ewell's left, which Jackson believed would threaten the right flank of Banks's column. With Colonel Charles Ronald's Stonewall Brigade in reserve, Winder moved Taliaferro's and Colonel Thomas Garnett's brigades forward in attack formation. Anticipating the direction of the main engagement, Jackson faced the brigades to the right. Doing so inadvertently exposed the Confederate left flank to two Federal brigades concealed in thick woods northwest of the Culpepper road. By midafternoon, Jackson's loosely strung front stretched from the northern edge of Cedar Mountain on the right to a heavy patch of woods on the left, with Hill's division strung along the Culpepper road in the rear.

Banks had about 9,000 men on the field to contend with about 21,000 Confederates from Jackson's corps. Banks confronted Ewell with three brigades from Major General Christopher Auger's division. Across the road on the right he placed two brigades from Major General Alpheus Williams's division. Two brigades commanded by George Gordon and Samuel W. Crawford remained concealed on the Federal right. Banks had additional brigades advancing on the road, and he also had authority from Pope to assume command of any other units coming up in the rear.

Although artillery exchanges occupied most of the afternoon, the fight intensified at 4:30 p.m. Scouts observed Federals massing in the woods on the Confederate left and warned Jackson. Winder suffered a mortal wound and was carried from the field after a shell sheared off his arm and tore open his side. Taliaferro assumed command of the division without having a clue as to Jackson's plans. Jackson had no time for a discussion and rode through Taliaferro's brigade to warn Garnett that Federals concealed in the woods were forming to strike the brigade's left flank. Taliaferro spotted the Federals and made a desperate attempt to pivot enough of his force to meet the threat.

At 5:00 p.m. Crawford's Federals, with bayonets fixed and regimental flags flying, struck Garnett's brigade. The Confederates fired a few hurried volleys and fell back on Taliaferro's brigade. Taliaferro had not completely faced his brigade about when Crawford struck. The entire left wing of the Confederate line collapsed, along with the Stonewall Brigade, and began to fall back on Hill.

At nearly the same moment, Augur's Federals flanked Early's brigade and threw it into disorder. An aide acting as an observer galloped across the field and warned Jackson that Early's flank was crumbling. Jackson grabbed a battle flag and rode toward the fight, shouting "Follow me!" He steadied the collapsing lines and then rode back to bring up Hill's 12,000-man division.

By force of sheer numbers and not by any notable tactics, Jackson's corps regained control of the battlefield and gradually pushed the Federals back. Brigadier General Lawrence O. Branch's brigade

from Hill's division struck Crawford's brigade and decimated it. Gordon's blueclads attempted to support Crawford but were also driven back when Brigadier General James J. Archer's Texas Brigade followed Branch into the fight. Late in the afternoon, Hill led the final charge of the day, and by 6:30 p.m. the enemy had been swept from the field.

Banks had managed the early stages of the battle of Cedar Mountain, also called Cedar Run and Slaughter Mountain, better than Jackson and probably missed a magnificent opportunity by failing to order up nearby reinforcements. Jackson first made the mistake of ignoring the threat on his left, and then made faulty dispositions. Jackson's biggest mistake was not providing his commanders with the information they needed in order to make adjustments while engaged in battle. Jackson's victory, though devoid of any particular brilliance, did disarrange Pope's plans and shifted all the momentum to General Lee. Pope also claimed victory, but bragging and blustering was just part of his personality.

The battle of Cedar Mountain came as a surprise when Jackson's corps collided with Bank's Federals at Cedar Creek, forcing the Confederates to rush into position at Slaughter Mountain. Both infantry and artillery make hasty arrangements, and not in the best field positions.

BATTLE OF CEDAR MOUNTAIN CASUALTIES

	KILLED	WOUNDED	MISSING	TOTAL
CONFEDERATE	231	1,107	0	1,338
FEDERAL	314	1,445	594	2,353

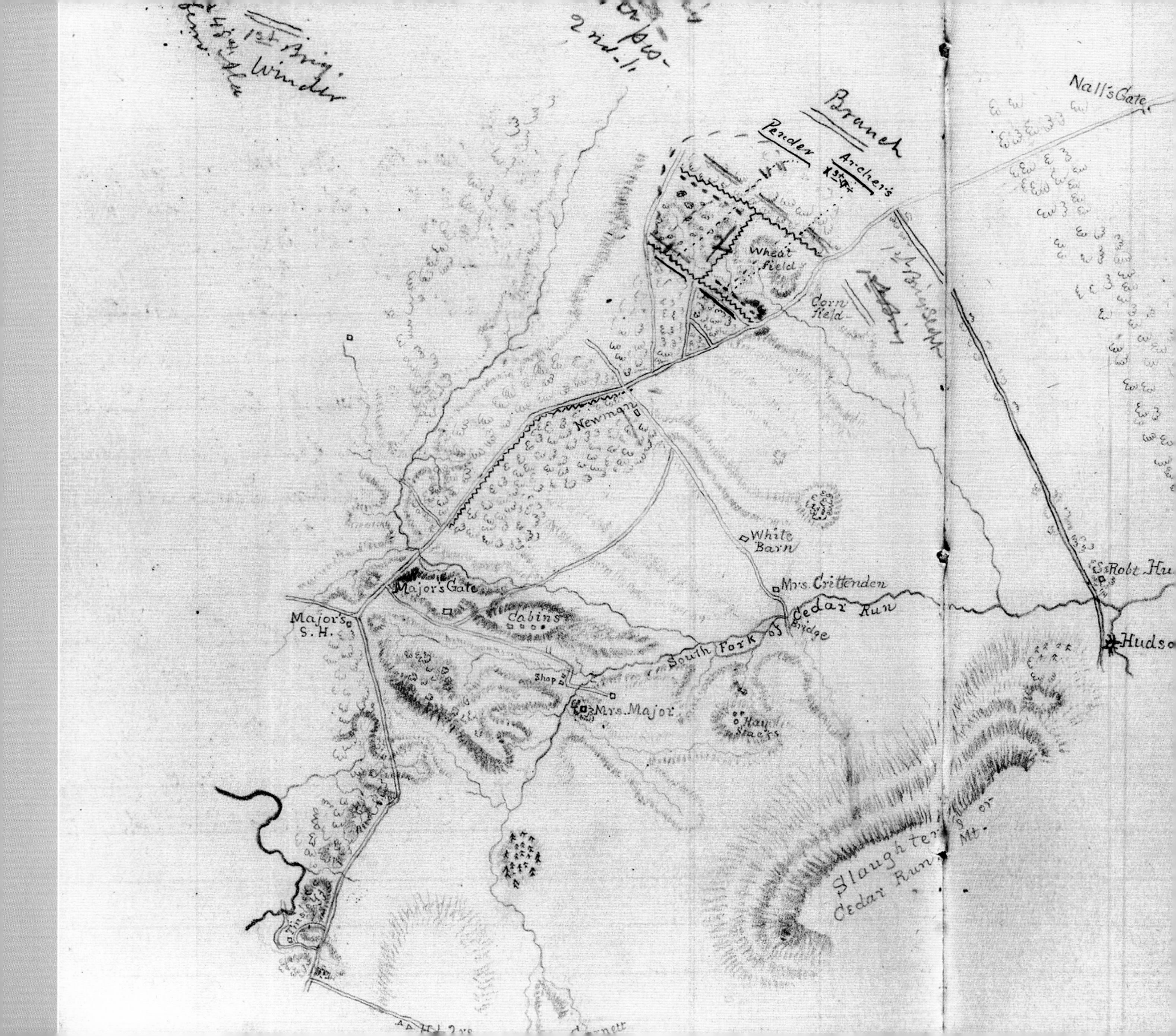

Nall's Gate
Branch
Pender
Archer's
Wheat field
Corn field
Newman
White Barn
Mrs. Crittenden
Cedar Run
Bridge
South Fork of
Major's Gate
Cabins
Major's S. H.
Shop
Mrs. Major
Hay Stacks
Robt Hu
Hudso
Slaughter's or Cedar Run Mt.

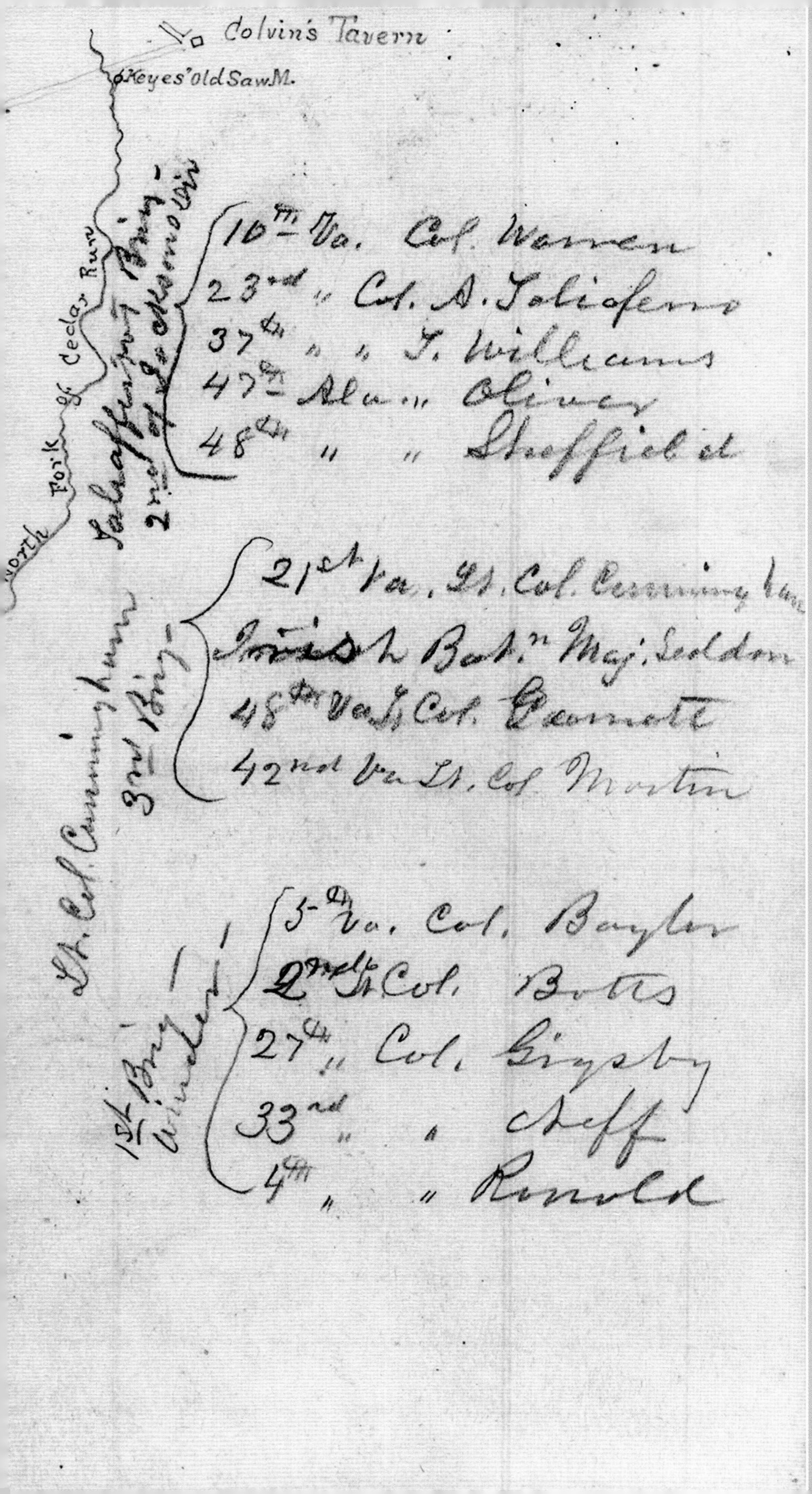

Colvin's Tavern

Keyes' Old Saw M.

North Fork of Cedar Run

Taliaferro's Brig. 2nd of Jackson's Div.

10th Va. Col. Warren
23rd " Col. A. Taliaferro
37th " " T. Williams
47th Ala " " Oliver
48th " " Sheffield

Lt. Col. Cunningham 3rd Brig.

21st Va. Lt. Col. Cunningham
Irish Batn. Maj. Seddon
48th Va. Lt. Col. Garnett
42nd Va. Lt. Col. Martin

1st Brig. Winder's

5th Va. Col. Baylor
2nd Lt. Col. Botts
27th " Col. Grigsby
33rd " " Neff
4th " " Ronald

Battle of Cedar Mountain Sketch [Map #5-20]

Hotchkiss began to prowl the field as soon as the action began to sketch roads and prominent terrain features. He noted some of the positions occupied by troops, and months later collected additional information that went into a comprehensive map (see page 53) that he provided for Jackson's official report.

A TWO–WEEK INTERREGNUM

Hotchkiss witnessed the battle without participating in the fighting, mainly because he worked as a civilian topographer. He often moved about the battlefield with no fear of being killed, or he stood on a hill to get a bird's-eye view of the action. Men fell around him while he watched the action, but Hotchkiss didn't even receive a scratch.

On the following afternoon he wrote, "We have occupied the same ground all day, holding the battle field, and have been busy bringing away our wounded, burying the dead, collecting arms, etc. I rode over the battle field and heard the most agonizing shrieks and groans from the many wounded Yankees that were lying in the full blaze of the sun." Hotchkiss did not have much time for sightseeing. After he finished sketches of the battlefield, Jackson ordered a new series of maps covering the routes and terrain features of an area extending from Gordonsville to Washington. "Do not be afraid to make too many," said the general, who then asked for five copies. Knowing Jackson's aversion to laboring on Sunday, Hotchkiss asked whether he should continue working the staff on the Sabbath. When the general answered in the affirmative, Hotchkiss noted in his journal, "We are all busy on the maps and all think the move is soon to take place."

Convinced that McClellan no longer posed a threat on the peninsula, General Lee began to move Longstreet's 30,000-man corps to Gordonsville, leaving only 25,000 men to protect Richmond. Lee anticipated greater resistance when he learned that McClellan was beginning to transfer some of his brigades to Washington. If the entire Army of the Potomac returned to Washington and assimilated Pope's army, Lee would be contending against a Union force of more than 140,000 blueclads.

On August 15 Pope controlled about 55,000 troops, but they were scattered about Virginia. Lee had a force of similar size, which he considered more than adequate to defeat Pope. Lee knew he had no time to waste.

When the Confederates entered the Union depot at Manassas Junction, they found tons of stores sitting and waiting for delivery to General Pope's army. Before burning the depot, Jackson's troops took the day off to plunder the stores and fill their empty knapsacks with ham and bacon, and their canteens with whiskey.

LEE MOVES NORTH

Unlike Jackson, Lee thoroughly discussed his battle plans. On arriving at Gordonsville, he established headquarters and called a meeting of his key officers and corps commanders. Jackson brought five sets of maps produced by Hotchkiss and passed them to the others. During this conference, the officers formed a bold plan. They would move the war north and away from Virginia's harvest, cripple Pope's army, and threaten Maryland and Pennsylvania. The movement called for Stuart's cavalry to demolish the railroad bridge over the Rappahannock, thus separating Pope from his source of supply. The plan also called for Jackson to conceal the movement of his corps by marching around Clark's Mountain east of the Orange and Alexandria Railroad, crossing the Rapidan, attacking Pope's left flank, and crushing the Army of Virginia before McClellan's reinforcements arrived. Lee would follow, moving north with Longstreet's corps to join Jackson at Manassas.

The complicated plan had a good chance of success until a detachment of Federal cavalry surprised Jeb Stuart, who lost his plumed hat while frantically making his escape. One of Stuart's staff officers was less fortunate. The Federals captured him along with his dispatch case, which contained Lee's plan. When Pope received the document, he hurriedly crossed the Rappahannock and began to move his army north. A few days later, Stuart raided Pope's headquarters and stole the general's dispatch case. Lee read the contents and noted that McClellan's troops were nearby. Lee's plan depended on swift marches, Pope's on reinforcements. Lee could have detained the Second Corps, but he put trust in Jackson's speed and cut him loose. On August 24 Jackson issued orders for his men to cook three days' rations, leave their knapsacks behind, fill their canteens, and be ready to march in the morning.

JOHN POPE (1822–92)

Born in Louisville, Kentucky, John Pope entered West Point at the age of sixteen and graduated in 1842, seventeenth in a class of fifty-six. Baretted captain during the Mexican War, Pope appeared to have a bright future ahead of him in the U.S. Army. A month after the outbreak of the Civil War, the war department boosted Pope to brigadier general and placed him in command of the Army of the Mississippi. He was promoted to major general on March 21, 1862, for distinguished service in Missouri, Tennessee, and Mississippi; the forty-year-old commander became President Lincoln's unanimous choice to lead the Army of Virginia.

Pope's command could not have gotten off to a worse start. On July 14, 1862, he boldly addressed the officers and enlisted men of the Army of Virginia by contrasting them unfavorably with Federals fighting in the West, where, he said, "We have always seen the backs of our enemies." He criticized the officers of his new command for "taking strong positions and holding them" and preparing "lines of retreat" as opposed to fighting aggressively. When asked by a reporter where his headquarters would be, Pope ostentatiously replied, "In the saddle." The quip prompted observers to suggest that Pope had put his headquarters where his hindquarters belonged.

After offending his men, Pope then angered Confederates by encouraging his soldiers to confiscate the property of Virginians and arrest any citizen suspected of aiding the enemy. Lee grew furious when he read the address. He vowed to destroy Pope—and did.

JAMES EWELL BROWN STUART (1833–64)

Born in Virginia in 1833, "Jeb" Stuart graduated from West Point in 1854, joined the U.S. Mounted Rifles, and prior to the Civil War spent the majority of his time on the frontier. On May 3, 1861, he resigned to join the Confederacy, although his father-in-law, Philip St. George Cooke, stayed with the Union. Stuart quickly transferred from the Confederate infantry to the cavalry and served with Stonewall Jackson during the spring of 1861 at Harpers Ferry. After performing conspicuously at First Manassas, Stuart became a brigadier general and spent the balance of his career with the Army of Northern Virginia.

During the Peninsula campaign, Stuart attracted enormous attention on June 12–15, 1862, by leading his brigade completely around the Army of the Potomac, which exposed weaknesses in General McClellan's dispositions. He again distinguished himself as a dashing cavalry leader in the Seven Days battles and at Harrison's Landing, which led to his promotion to major general on July 25, 1862, and command of the Army of Northern Virginia Cavalry Corps. Stuart's consistently brilliant exploits made him one of the fabled cavalrymen of the Civil War, but his carelessness and lapses in discipline sometimes annoyed General Lee.

Stuart's West Point classmates called him "Beauty" for his massive beard, which some claimed covered a receding chin and camouflaged his youth. He stood about five feet nine inches with a solid, nearly square build. When rollicking through the countryside, he often stopped to woo the ladies and have "his man" strike up the banjo. Stuart loved life and exhibitionism, though he was both religious and a serious professional. Lee called Stuart the "eyes of the army." Despite an occasional lapse in judgment, Lee said his cavalry commander "never brought me a piece of false information." When Stuart died after being shot at Yellow Tavern in May 1864, Lee could find no one to replace him.

THE FOOT CAVALRY MARCHES NORTH

Before dawn on August 25, Jackson put his rested column in motion. At 8:00 a.m. Federal signal stations observed a long cloud of choking dust, counted the number of regimental flags, and informed Pope. Despite having a copy of Lee's orders, Pope expected Jackson to return to the Shenandoah Valley. It did not occur to Pope that Jackson might be maneuvering to strike the Federal rear.

Using maps prepared by Hotchkiss, whom Jackson ordered to remain in camp to complete a new project, the Confederates followed roads scouted and chosen by Captain Keith Boswell who, as Jackson's chief engineer, worked closely with Hotchkiss in developing the route. Men dropped from fatigue and thirst in the torrid heat. At Salem, where Jackson's column turned east paralleling the Manassas Gap Railroad, the veteran infantry now knew where they were going. Ahead were the Bull Run Mountains and Thoroughfare Gap, which led to Manassas Junction and Pope's massive supply depot. A question rippled through the column: Would Pope's army be waiting at the gap? Jackson pressed forward and found the gap undefended without so much as a Federal sentry on duty.

At midmorning on August 26, the Confederate column reached Gainesville, twenty miles behind Pope's line on the Rappahannock, and turned southeast toward Bristoe Station and the Orange and Alexandria Railroad. Details cut tracks and burned the trestle over Broad Run, completely severing Pope from his supplies. After marching fifty-four miles in two days, the famished foot cavalry arrived at Manassas Junction after midnight on August 27 and overwhelmed the guards. Personal hunger took precedence over Jackson's orders to burn the depot, destroy the whiskey, and move on. The general paused while his men looted the depot, collected vast stores of rations, and then torched what was left.

Lee expected the Second Corps to take a favorable position on the old Bull Run battlefield and wait for the rest of the army to arrive. Jackson expected to be attacked and used his initiative when, on August 28, a Federal column appeared on the pike along the railroad cut. The fighting began that evening and steadily intensified over the next two days.

A FAMILIAR FIELD OF BATTLE

Jackson's knowledge of the area stemmed from his participation at First Manassas. On the night of August 27, he moved his corps to Stony Ridge (Sudley Mountain) and began to spread his brigades along the recess of an unfinished railroad about a mile northwest of the village of Groveton. Before moving all the brigades into position, he attempted to mislead and mystify the Federals by sending A. P. Hill's division toward Centreville. Ewell moved along the north side of Bull Run as if to support Hill, and Taliaferro moved directly across the Sudley Road to the Warrenton turnpike. By midafternoon on August 28, Jackson had his entire corps in position along the unfinished railroad cut, with cavalry and artillery placed on the flanks.

Pope rushed forces north without having any plan of operation. He completely ignored Longstreet. Federals entered the smoldering depot of Manassas on the morning of August 28 and found Jackson gone. Pope heard reports of Confederates at Centreville, assumed he had discovered Jackson's point of attack, and moved his entire force toward the town. Baffled by Jackson's maneuvering, Pope soon learned that a division led by Brigadier General Rufus King, which had been moving westward along the Warrenton Turnpike, clashed with a Confederate force near Groveton. Pope reversed the direction of his movement and around dusk sent the blueclads across the old Bull Run battlefield with instructions to assemble at Groveton, where King's brigade had encountered Confederate skirmishers. Pope did not find Jackson's corps waiting at Groveton, which troubled him. King's skirmish with Confederates at Groveton on August 28, however, set the stage for the forthcoming battle.

During the battle of Groveton, Federal forces advanced on Jackson's artillery firing from the railroad cut in the distance. General Pope had no idea he was leading his army into a well-organized trap. Hotchkiss was not present to witness the battle, but spent several days afterward mapping the battlefield.

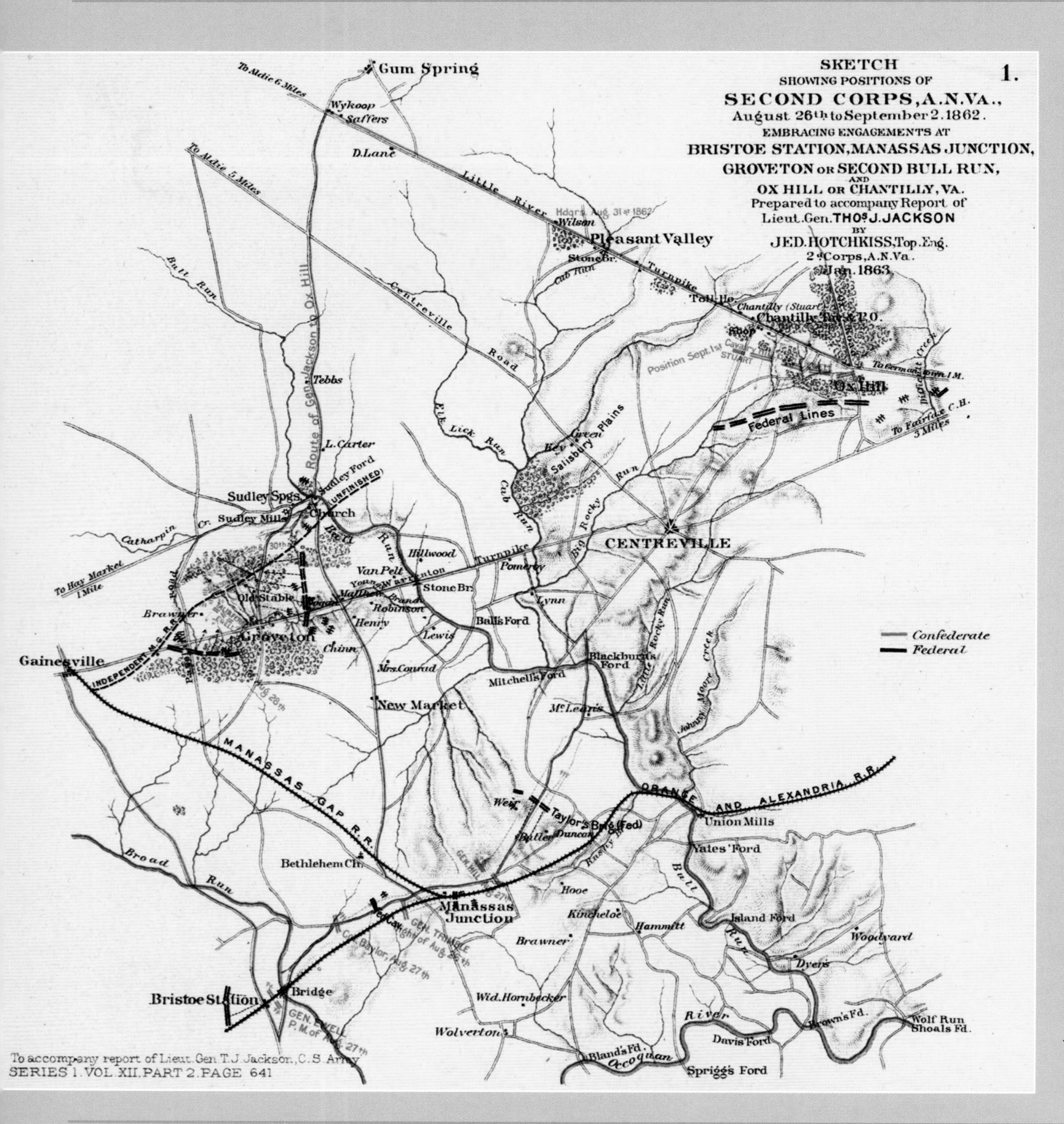

Second Manassas Battlefield [Envelope Map #109]
The original map sketched by Hotchkiss of the Second Manassas (Second Bull Run) battlefield became the base map from which Hotchkiss prepared and refined a second map (see below), similar to the one General Jackson sent to General Lee on April 27, 1863. The two maps together demonstrate the manner in which Hotchkiss first organized a map from sketches and then produced a finished map as time and the collection of data allowed.

Second Manassas Battlefield [Map #111-1]
Hotchkiss arrived on the battlefield after the action ended. Knowing that Jackson would want a map to submit with his battle report, Hotchkiss immediately began sketching the entire battlefield as it pertained to the positions occupied by Jackson's II Corps. The finished map, as submitted by Jackson, became part of the Official Records.

SECOND MANASSAS (SECOND BATTLE OF BULL RUN): AUGUST 28–30, 1862

Second Manassas, also called the second battle of Bull Run, turned from a skirmish into a full-scale fight on the afternoon of August 29 when Major General Samuel P. Heintzelman's Third Corps and Major General Jesse L. Reno's Ninth Corps attacked Jackson's Second Corps massed behind the embankment of the unfinished Manassas Gap Independent Railroad. The Federal assaults were piecemeal and poorly coordinated. A few well-directed Confederate volleys sent the Federals reeling back in disorder.

Jackson's original orders indicated that he was to take a position on the enemy's flank strong enough to hold back Pope's army until Lee arrived with Longstreet's corps. Jackson also wanted a flexible position from which he could attack or, if Pope blocked Longstreet, retreat. Instead, he offered battle on August 28 before Longstreet arrived. Most military authorities would question Jackson's eagerness to challenge Pope's reinforced 62,000-man army with only 20,000 of his own, but Jackson held a low opinion of Pope's military skills and expected Longstreet to arrive around noon.

The first regiments from Longstreet's corps began to move into position on Jackson's right flank at 11:00 a.m. Much to Jackson's surprise, Longstreet put his force into position but refused to attack, thus depriving Lee of an opportunity to destroy the bulk of Pope's army. At the time, neither side recognized that a two-mile gap existed on the Confederate left, which could have been turned by Major General Fitz-John Porter's division laying idle a few miles to the south, waiting for orders from Pope. Porter's corps represented the first of McClellan's army to arrive on the field and, had he attacked forcibly, would have caused considerable chaos among the Confederates.

Jackson counterattacked from his defensive position late in the afternoon, fully expecting Longstreet to join the assault. When Longstreet did nothing and remained on the defensive, Jackson withdrew to the railroad cut after exhausting much of his ammunition. Pope interpreted Jackson's retirement as a retreat and ordered a vigorous pursuit in the morning.

Pope's assault the next day did not begin until 3:00 p.m. Jackson withstood the first attack, inflicting heavy casualties on the Federals. The Union commander still failed to comprehend that Longstreet's corps occupied the right of the Confederate line, partly because Longstreet had taken a defensive position and let Jackson do the fighting. At 4:00 p.m. Lee took advantage of an opportunity by letting Pope throw his entire army against Jackson's extended front. As the Federals advanced and engaged the Confederate left, Longstreet's corps came pouring out of the woods and down the slopes from the Confederate right, tearing into Pope's weak left flank and routing three Federal brigades. Pope had enough manpower to form a defensive position on Henry House Hill, which enabled most of his force to flee across the Stone Bridge and into Centreville.

Lee attempted to maintain the offensive on September 1 by sending Jackson's corps around Centreville to flank Pope. Jackson's well-traveled foot soldiers struck two Federal divisions commanded by Brigadier General Isaac I. Stevens and Major General Philip Kearny at Chantilly (Ox Hill). The encounter lasted until midnight and resulted in both Union commanders being killed. Although Pope had thousands of reinforcements nearby, he withdrew to the defenses of Washington.

SECOND MANASSAS CASUALTIES, INCLUDING CHANTILLY

	ENGAGED	KILLED	WOUNDED	MISSING	TOTAL
CONFEDERATE	48,527	1,481	7,627	89	9,197
FEDERAL	75,696	1,724	8,372	958	11,054

HOTCHKISS'S RETURN

On August 26, after working for two days on maps, Hotchkiss packed his instruments and sketchbooks and followed in the wake of Longstreet's column. He obtained enough news of the fighting at Manassas to insert a running account of it in his journal. Hotchkiss finally reached Groveton on August 31 and spent the day riding over the battlefield, "where they are burying the dead and collecting the wounded . . . sometimes in heaps." Among the dead lay Colonel William Baylor of the Fifth Virginia, the man responsible for getting Hotchkiss assigned to Jackson's staff.

Summoned by Jackson the following morning, Hotchkiss spent about thirty minutes going over maps with the general. Jackson seemed satisfied, instructed Hotchkiss to make a thorough but rapid survey of the battlefield, and then departed with his corps to fight a second battle near Chantilly. Hotchkiss had barely begun to collect information for the general's map when he received a summons from Jackson on September 3. Lee had issued orders to move north, and Jackson tasked Hotchkiss with leading the army. With the mapmaker in the van, the Army of Northern Virginia marched to Dranesville and reached Leesburg the following day. With only old charts for reference, Hotchkiss had his staff spread out so they could add roads, streams, hills, woods, and farms to their base map.

On September 5 Jackson's corps crossed the Potomac at White's Ford and marched into Maryland. The men cheered, and when a band struck up the tune "Maryland, My Maryland," thousands of voices picked up the lyrics and sang as they splashed across the Potomac. The footsore troops purchased melons from a barge on the river, and General Jackson purchased an entire field of ripe corn so the men could roast ears and fill their knapsacks.

LEE MOVES WEST

While the army rested for several days at Frederick, Hotchkiss raised his tent near headquarters and prepared a series of new maps requested by Jackson. Several times a day, Jackson called for maps. Hotchkiss usually found the general in conference with Lee, Longstreet, and Stuart. He seldom stayed after answering a few questions and sometimes returned to his tent to comply with requests for more maps.

At 3:00 a.m. on September 10, the entire army marched from Frederick and headed west. Hotchkiss did not know until later that General Lee had split the army at South Mountain, sending Jackson back into the lower Shenandoah Valley to clear the Federals out of Martinsburg and Harpers Ferry. Lee hoped to move north into Pennsylvania, perhaps capture Harrisburg, the state's capital, and sever the Pennsylvania Railroad by destroying its bridges. Before entering Pennsylvania, Lee saw the logistical necessity of restoring communications in the Shenandoah Valley, through which he would be forced to draw supplies. He detached Jackson's corps to complete the all-important strategic mission while retaining most of Longstreet's corps in Maryland.

In early September, President Lincoln transferred Pope's defeated divisions to the Army of the Potomac and ordered General McClellan to pursue the Army of Northern Virginia and crush it. McClellan reached Frederick a few days later, found the Confederates gone, and pitched tents on the enemy's former campground. A private wandering through camp observed a small package and inside found three cigars wrapped in Lee's Special Order No. 191, dated September 9, 1862. Lee had distributed seven copies of the order, something Jackson never did, and for the second time in less than a month the Federals obtained complete information on Lee's planned movements. With a force twice the size of Lee's, McClellan now had the chance of redeeming his apathetic performance on the peninsula by trapping Lee's divided army and possibly ending the war. "Here is a paper," proclaimed McClellan to his staff, "with which if I cannot whip Bobbie Lee I will be willing to go home."

THE BATTLE OF CRAMPTON'S GAP: SEPTEMBER 14, 1862

With Jackson's corps on a separate route to Harpers Ferry, Lee moved west with Longstreet's corps without realizing that McClellan's entire army was in pursuit. McClellan might have been more aggressive had General Halleck not warned the general of the possibility of being baited into a trap. McClellan, however, moved at uncustomary speed and around noon on September 14 struck two small cavalry and three small Virginia regiments under Colonel William A. Parham at Crampton's Gap. With his army diluted because of the absence of Jackson's corps, Lee had distributed most of Longstreet's force at the gaps of South Mountain. He had sent Major General Lafayette McLaws's division and that of Major General Richard H. Anderson to Maryland Heights. Both divisions were engaged in dragging artillery up and through the tangled woods and ravines of the mountain to support Jackson's assault on Harpers Ferry. Threatened from the rear, McLaws had to turn his own infantry and part of Anderson's on Union Major General William B. Franklin's 12,000-man Twelfth Corps as it pressed into Crampton's Gap against Parham's weak force.

Parham's regiments tenaciously held the pass for several hours, holding off a Federal division commanded by Major General Henry W. Slocum. Franklin then sent Major General William F. "Baldy" Smith's division into the gap, and Parham's resistance weakened under mounting Union pressure. Brigadier General Howell Cobb arrived with four fresh Confederate regiments but could not stem the Federal assault. Under more severe pressure from Franklin's corps, McLaws detached two brigades and two cannons from Maryland Heights to support Cobb.

Franklin eventually forced most of McLaws's and Anderson's infantry out of the pass and into a poorly defensible position in Pleasant Valley. A concerted Federal attack down Pleasant Valley could have destroyed McLaws's division. Although Franklin had gained the advantage, he overestimated McLaws's strength, lost his nerve, and doomed Harpers Ferry.

THE BATTLE OF SOUTH MOUNTAIN (TURNER'S GAP): SEPTEMBER 14, 1862

On the morning of September 14, Brigadier General Alfred Pleasonton's Federal cavalry began to probe Turner's Gap at the northern end of South Mountain about the same time Franklin assaulted Confederates defending Crampton's Gap. South of Turner's Gap, a winding road snaked through the much smaller pass at Fox's Gap. Both gaps and a number of trails across South Mountain were points that required defensive measures.

After Pleasonton's cavalry had failed to penetrate Turner's Gap, Brigadier General Jacob D. Cox arrived with a division of Federal infantry at 9:00 a.m. On approaching Turner's Gap, he struck Brigadier General Samuel Garland's brigade of North Carolinians, the only brigade on hand from Major General Daniel Harvey Hill's division. As the battle intensified, Hill sent in another brigade of North Carolinians commanded by Brigadier General George B. Anderson. General Burnside, who was commanding McClellan's left wing, responded by sending Major General Jesse Reno's Ninth Corps against Garland's brigade at Fox's Gap. In a jumbled battle over rugged terrain, both Reno and Garland were killed. The remainder of Reno's corps arrived on the field at noon, but for two hours failed to attack. During the interregnum, Hill called up his remaining two brigades, which consisted of Brigadier General Robert Rodes's Alabamans and Brigadier General Roswell S. Ripley's Georgians. Not far behind Rodes and Ripley were two Confederate divisions, one under Major General David R. Jones and the other under Brigadier General John B. Hood. Both divisions marched hurriedly across dust-choked roads toward the sound of battle.

Late in the afternoon, McClellan added Major General Joseph Hooker's First Corps to the constantly intensifying battle at Turner's Gap. Federals charged up the slope and fell to Confederate volleys fired from knolls and wooded stretches on the mountainside. Jones's and Hood's divisions ascended South Mountain and added considerably more firepower to the defense of the gaps. The fight continued into the night, and at 10:00 p.m. the Confederates retired from the mountain and yielded the high ground covering Turner's Gap. Lee now had a serious problem and withdrew to Antietam Creek to concentrate his scattered forces.

Hotchkiss never made maps of the crucial battles at Crampton's Gap and South Mountain because he was at Harpers Ferry with General Jackson.

Major General Daniel Harvey Hill (1821–89)
In the days ahead, Lee realized that McClellan had found a copy of his special orders. The only missing copy belonged to General Hill, who blamed one of Lee's staff officers for losing the document and claimed the orders came to him through another source. Hill smoked cigars, leaving posterity to decide who was telling the truth.

BATTLE OF SOUTH MOUNTAIN CASUALTIES

	ENGAGED	KILLED	WOUNDED	MISSING	TOTAL
CONFEDERATE	17,852	352	1,560	800	2,712
FEDERAL	28,480	325	1,403	85	1,813

JACKSON AT HARPERS FERRY

On September 9, 1862, Jackson began a long, circuitous march to Harpers Ferry, following routes provided by Hotchkiss. Crossing the Potomac at Williamsport, Confederates chased Brigadier General Julius White's brigade out of Martinsburg and sent the blueclads fleeing into Harpers Ferry. Having opened the Valley Turnpike, Jackson turned east toward Harpers Ferry at the same time that Major General Lafayette McLaws's and Brigadier General John G. Walker's columns from Longstreet's corps were placing artillery on Maryland Heights and Loudoun Heights, two crests looming high above Harpers Ferry. McLaws's division was also engaged at the same time with Franklin's corps at Crampton's Gap, and the firing could be heard at Harpers Ferry.

Jackson wanted to coordinate his assault on the town and sent Hotchkiss across the Shenandoah to confer with Walker on Loudoun Heights and to arrange an exchange of signals. Hotchkiss climbed the mountain, made the necessary arrangements, and watched from the crest as the battle took shape.

Major General Henry Halleck, Lincoln's general-in-chief, made an enormous blunder by placing 12,000 Federals at Harpers Ferry. Having been there in 1861, Jackson knew the town could not be defended. Halleck exacerbated the blunder by leaving fifty-seven-year-old Colonel Dixon S. Miles, a man prone to intoxication and indecisiveness, in charge. When General White arrived with his brigade from Martinsburg, he allowed Miles to discourage him from taking overall command and watched the catastrophe develop.

Early on September 15, Jackson's artillery opened from Camp Hill, a village on a rise overlooking Harpers Ferry. Immediately afterward, artillery shells from McLaws's and Walker's batteries began crashing into Federal defensive works. Colonel Miles immediately surrendered and, while dashing about the field ordering his men to give up, suffered a mortal wound from a bursting artillery shell. General White, who had abrogated his command responsibilities to Miles, now assumed the embarrassing chore of surrendering the town. Hotchkiss came across the Shenandoah after the surrender and "went among the prisoners and conversed with them," noting that "most of them were only two months from home and were glad they were taken."

Meanwhile Lee, after fighting drawn battles on September 14 at Crampton's Gap and South Mountain, learned that McClellan knew his plans. He instantly sent urgent orders to Jackson to return to Maryland. Jackson received the message the morning he assaulted Harpers Ferry. He reviewed a map with Hotchkiss and left A. P. Hill's division behind to parole 12,000 prisoners and collect 13,000 arms, seventy-three cannons, and tons of badly needed supplies. He then pressed forward with two divisions, reaching Shepherdstown near Sharpsburg that night.

Hotchkiss remained at Harpers Ferry because his fatigued horse needed rest. Jackson never asked for a battle map, so Hotchkiss did not make one. Instead, Hotchkiss helped Hill's rear guard burn the railroad bridge over the Potomac and the pontoon bridge over the Shenandoah. On the evening of September 16, Hill marched out of Harpers Ferry and headed for Shepherdstown. Hotchkiss spent another night in the area resting his horse, noting in his journal, "We have removed most of the stores we captured, and the owners have claimed most of the slaves that we found in Harper's Ferry." Slaves lived miserably during the Civil War, being freed one day and then enslaved again the next.

Battle of Antietam: Opening Action [Map #116]

Hotchkiss believed that Jackson would eventually want a map of the action at Antietam. He made several rough sketches to capture the terrain without attempting to show the positioning of the Confederate and Union forces. Hotchkiss's normal routine was to take his collection of sketches and build a comprehensive map of the battlefield, which is in the envelope, and then attempt to show the positioning of the various units engaged in the fight, which for Hotchkiss became an impossible task. He must have been relieved when Jackson never asked for an after-action map.

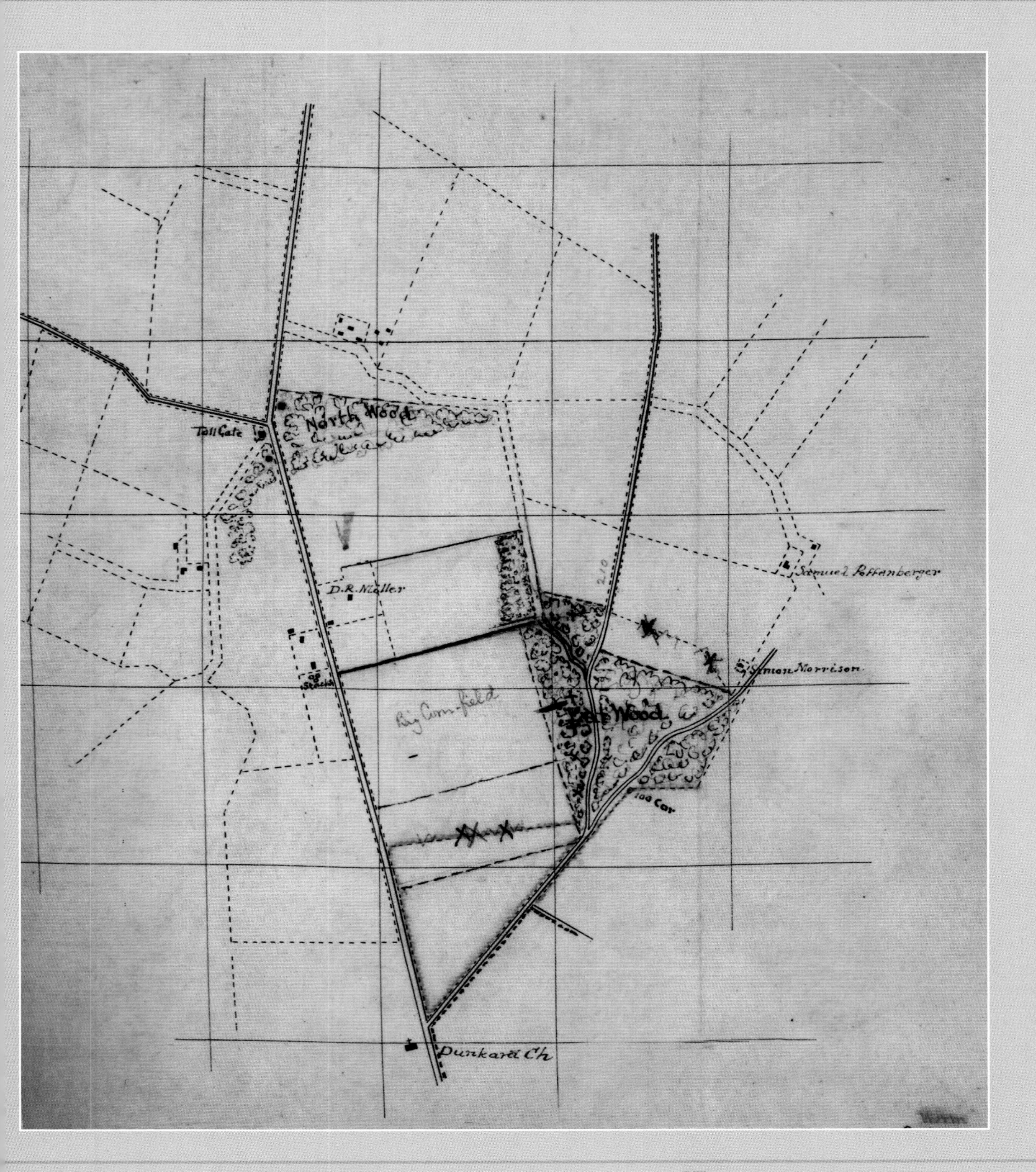

Battle of Antietam
[Envelope Map #113]

Hotchkiss produced a large map of the Antietam battlefield long after the engagement ended. On September 17, 1862, as the battle roared across the cornfields and meadows of the quiet farming community of Sharpsburg, Hotchkiss was not there but on the road from Harpers Ferry. He made several hurried sketches of the area, and in particular where Jackson positioned his Second Corps. One of those sketches as seen here, shows where the battle opened in the North Woods on the Confederate left flank. Although Hotchkiss eventually completed a map that showed all the features of the battlefield, Jackson never asked that it detail the fight, perhaps because the only way to illustrate the attacks and counterattacks required a series of maps that covered a fourteen-hour engagement involving 130,000 men.

THE BATTLE OF ANTIETAM (SHARPSBURG): SEPTEMBER 17, 1862

On September 15, with about half of his army at Harpers Ferry and the other half congregated along Antietam Creek, General Lee made the decision to go on the defensive and reconcentrate his army around Sharpsburg, Maryland. He considered removing his forces from the area, but news of Jackson's victory at Harpers Ferry combined with McClellan's habitual caution changed his mind. For reasons not comprehensible to military authorities, McClellan did not press the action on the following day, thereby enabling Lee to throw up breastworks and significantly improve his defenses. Lee felt even more gratified when Jackson arrived with the Second Corps less Hill's division, which had remained at Harpers Ferry to parole 12,000 prisoners and collect the loot. Despite Jackson's arrival, a huge disparity in numbers existed. Lee's 40,000-man army, which had been reduced by straggling, should not have been difficult for McClellan's 75,000-man army to defeat.

At dawn September 17, McClellan committed the first of several errors by assaulting Lee's left flank in the first of a series of piecemeal attacks. Hooker's First Corps, followed later by Major General Joseph K. F. Mansfield's Twelfth Corps, came pouring out of the North Woods with a resounding cheer and plunged into a forty-acre cornfield. Artillery fire from Jackson's batteries, augmented by the staccato of blazing muskets, slowed down the Federal assault but did not stop it. The Federals gained the cornfield and drove Jones's and Lawton's divisions across the Hagerstown Pike and toward the tiny, whitewashed Dunkard Church at the edge of the West Woods. Jackson let Hooker's battle-fatigued veterans edge closer before turning Hood's brigade loose on the enemy's flank. Hooker's attack collapsed, and his corps staggered back through the shot-riddled cornfield to take refuge in the North Woods.

As Hooker's corps fell back, Mansfield's Twelfth Corps charged from the East Woods. The Federals swept across the cornfield, crossed the Hagerstown Pike, and struck Jackson's three divisions, which had just been mangled by Hooker's assault. Hundreds more fell, including Mansfield, who was killed, and Hooker, who suffered a wound.

McClellan had planned to penetrate Lee's flank and roll up the entire Confederate line from left to right, but neither Hooker's nor Mansfield's corps wedged a breakthrough. When Lawton fell back to reorganize, Jackson threw in reinforcements from Longstreet's corps. Three divisions, commanded by McLaws, Walker, and Anderson, began to fill voids in Jackson's defensive perimeter just in time to stop the assault of veterans from Major General Edwin V. Sumner's Second Corps, which at 9:00 a.m. came storming down a road that passed through the edge of the East Woods. Part of Sumner's attack crossed the Hagerstown Pike and passed into the West Woods. Walker's and McLaws's divisions and remnants from Jackson's three divisions raked the Federals from three directions. Sumner withdrew thirty minutes later, leaving 2,200 Federal casualties on the field. Jackson believed the momentum had changed and counterattacked, only to be thrown back with equally heavy casualties. He pulled back his forces to reorganize just as McClellan sent fresh divisions forward to assault the Confederate center.

THE CONFEDERATE CENTER AT ANTIETAM

At 10:00 a.m, believing Lee had pulled most of his force from the center to reinforce Jackson, McClellan sent Fitz-John Porter's Fifth Corps against Hill's Confederate line holding a sunken farm lane, an area that those who fought and survived called "Bloody Lane." McClellan then changed his mind and held most of Porter's force in reserve. As the flow of battle shifted to the Confederate center, Hill's thin

BATTLE OF ANTIETAM CASUALTIES

	ENGAGED	KILLED	WOUNDED	MISSING	TOTAL
CONFEDERATE	51,844	2,700	9,024	2,000	13,724
FEDERAL	75,316	2,108	9,549	753	12,410

line, backed by several batteries of artillery, stubbornly resisted despite having suffered severe losses at South Mountain. Jackson supported Hill's left flank with Richard Anderson's division. Sumner still had two available divisions and assaulted the left of Bloody Lane while some of Porter's brigades struck the center. When it appeared that pressure along Jackson's own left flank had subsided, Anderson sent more brigades into the sunken road to repel Sumner's assault. For three hours, Porter and Sumner hammered Bloody Lane without dislodging the Confederates.

Through misunderstood orders, a gap inadvertently opened on the Confederate line. For about twenty minutes, Lee's center lay exposed. Federals might have penetrated Bloody Lane, but McClellan committed another crucial error by refusing to release Franklin's Sixth Corps, which waited in reserve for just such an opportunity. The Confederates plugged the gap, counterattacked, and threw the Federals back. McClellan would later claim that General Burnside's wing should have struck the Confederate right, and had Burnside done so at any time during the course of the morning or early afternoon, the Confederate line could not have held.

THE CONFEDERATE RIGHT AT ANTIETAM

By the early afternoon, Lee had drained most of his resources while defending the Confederate left flank and center. The entire left wing of McClellan's army, commanded by General Burnside, had not crossed the stone-arched bridge on lower Antietam Creek because a small Confederate brigade led by General Toombs held back Brigadier General Jacob D. Cox's three divisions from the Ninth Corps, which had formed on the far side of the bridge. Despite persistent orders throughout the morning from

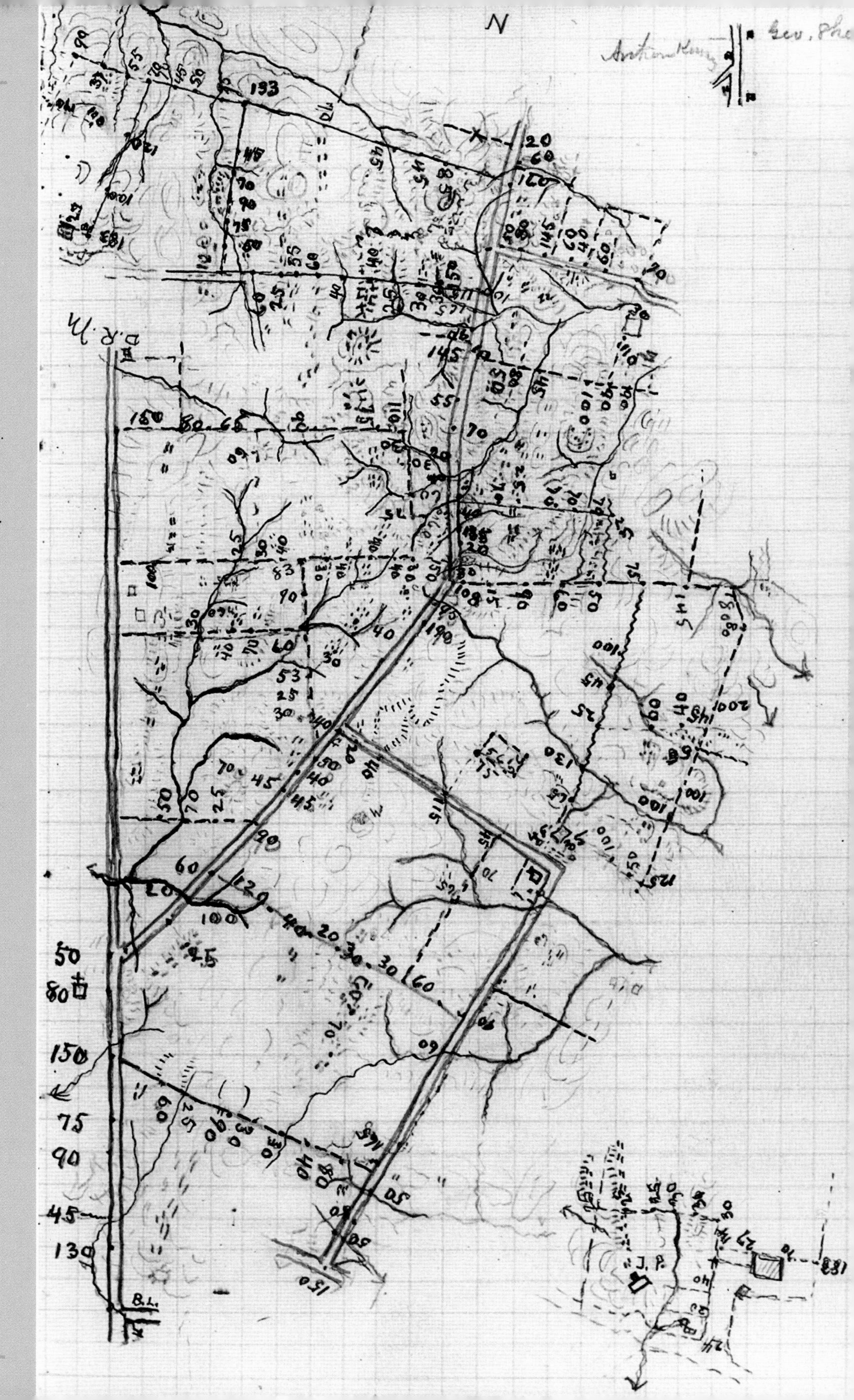

Antietam Battlefield [Sketchbook Map #4-4]
In 1864 Hotchkiss returned to the Antietam battlefield and spent time with his two assistants surveying the terrain. The sketch shows roads, streams, hills, elevations, and other details, such as the North Woods at the top of the sketch. The sketch serves as an example of the detail Hotchkiss developed before making some of his maps.

McClellan to push across the bridge and assault the Confederate right, Burnside could not budge Cox, who on several occasions started Federals across the bridge only to have them repulsed by Toombs's grayclads.

Finally at 3:00 p.m., Cox pushed Brigadier General Edward Ferrero's brigade across the bridge and, followed by the rest of the Ninth Corps, shoved Toombs aside and attacked the weak Confederate right at 4:30 p.m. Lee moved four exhausted brigades from the center and attempted to hold the Federals back. The blueclads, however, kept coming, pushing the Confederates to the outskirts of Sharpsburg.

A. P. Hill had just about finished his business at Harpers Ferry when, at 6:30 a.m., he received an urgent order from Lee to go quickly to Sharpsburg. Hill's Light Division was the largest in the Army of Northern Virginia and also one of the best. Hill put on his red flannel battle shirt, left one brigade at Harpers Ferry to finish gathering captured supplies, and put five brigades on the road to Sharpsburg. Hill's men marched eighteen miles in about seven hours, and the first brigade reached Lee's right flank just as Cox's corps began to re-form after crossing the stone span that derisively became known as "Burnside's Bridge." Suddenly from the south, Hill's brigades stuck Burnside's flank, driving the Federals back to the bridge. By dusk, the Confederates had cleared the battlefield of Federals, which apart from scattered units had returned to the far side of Antietam Creek.

ANTIETAM AFTERMATH

After "the bloodiest single day of the war," the strength of both armies had been sapped. Despite heavy casualties, Lee wanted to counterattack, but Jackson and Longstreet prevailed, insisting it could not be done. McClellan never considered attacking, even though nearly 24,000 of Porter's and Franklin's troops had not fired a shot the entire day and another 12,000 fresh troops sent by Secretary of War Stanton arrived on the morning of September 18.

Lee thumbed his nose at McClellan throughout September 18 and withdrew that night into Virginia. McClellan claimed victory because the Federals took possession of the battlefield. Antietam provided the springboard from which President Lincoln issued the Emancipation Proclamation. He generously gave McClellan praise for defeating Lee, though the battle was a drawn affair and not a clear victory for either side. After complimenting McClellan for winning the battle, Lincoln spent weeks of frustration trying to goad the general into pursuing Lee and ending the war. McClellan continued to stall. In his view, he had completed his mission by driving Lee out of Maryland and back into Virginia.

THE RETREAT OF LEE'S ARMY

When A. P. Hill's division departed from Harpers Ferry on the morning of September 17, Hotchkiss packed his maps and instruments, saddled up his tired horse, and followed. By the time he reached Shepherdstown, the fighting had abated. He found a place to sleep, and in the morning crossed the Potomac to Sharpsburg. He could not locate Jackson, but he was under the impression that the general wanted a reconnaissance conducted up the Potomac, above Shepherdstown. Hotchkiss saddled his horse, obtained a guide, and began scouting the fords suitable for crossing the army's wagons and artillery. He sketched the location of fords and bridges and the roads that led to them. If he brushed against enemy pickets, he also marked their location.

Having been accompanied by a small detachment of cavalry, Hotchkiss sent regular reports back to Sharpsburg. He returned later during the evening and found that "the road was full of horses, wagons, ambulances, etc., and very narrow, with a bluff above and below and the river beneath." He still had not located Jackson when Jeb Stuart approached and said that the general had seen the reports and that the army was withdrawing along the designated routes. Stuart asked Hotchkiss to guide him because Jackson and Lee had departed from Sharpsburg without leaving any instructions for the cavalry. The only sketch Hotchkiss salvaged from his reconnaissance is the one shown opposite. All the others had been forwarded to Jackson.

On Friday, September 19, Hotchkiss noted in his journal: "Our troops, trains, artillery, etc., were all safely over the Potomac by 8 a.m., having been covered by the fog, and the enemy did not find that we were all over [across the river] until everything had gained the Va. bluff, when [the Federals] came on rapidly and commenced shelling."

McClellan's failure to pursue allowed Lee's army to move south at a leisurely pace, slowly healing, and without being in the least molested.

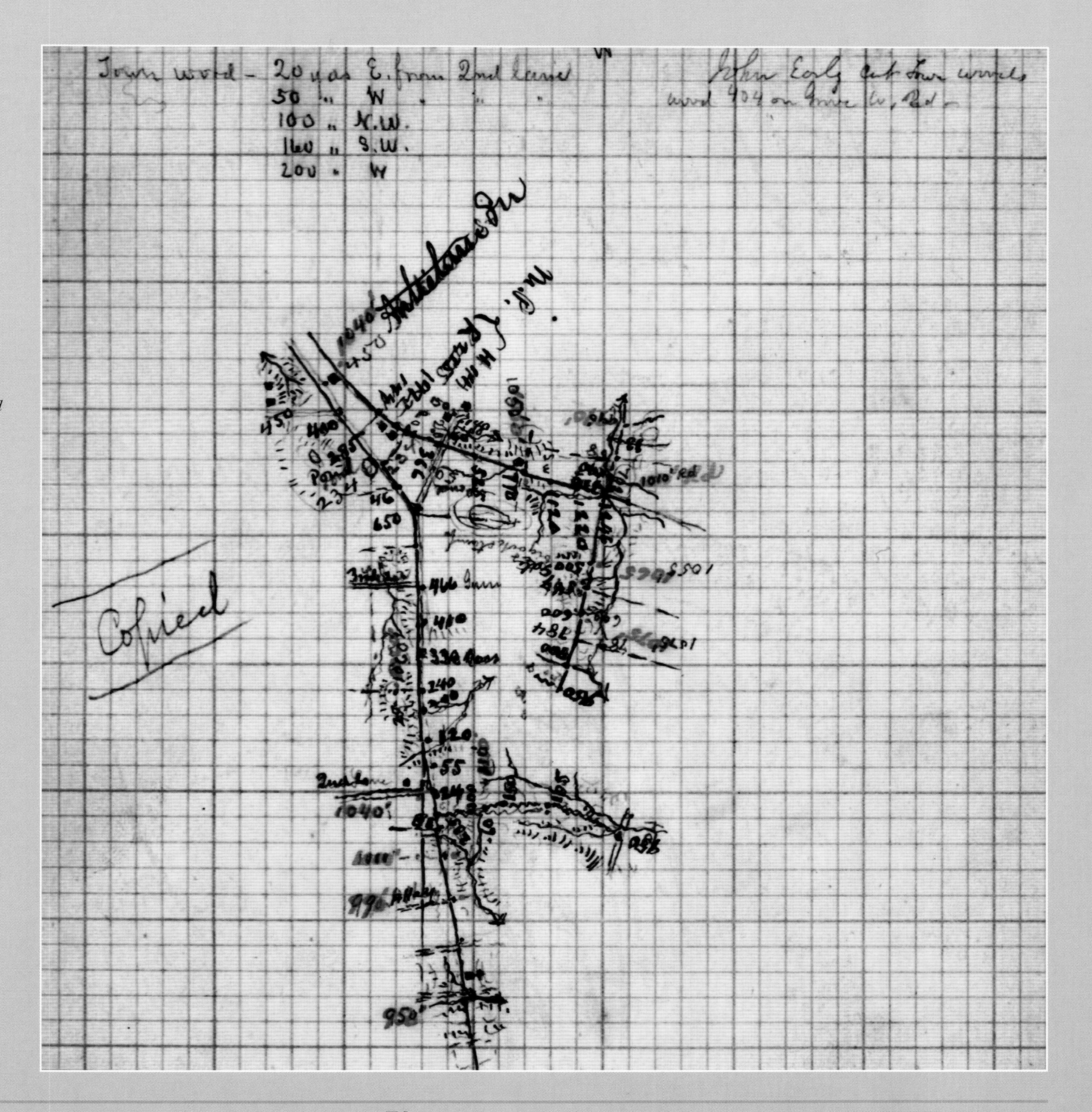

Escape Route Sketch
[Map #119]
On September 18 Jackson ordered Hotchkiss to scout the terrain around the Potomac River and develop escape routes for Lee's army. This sketch was one of several that Hotchkiss sent back to Jackson. It shows terrain features, roads, and distances between certain landmarks. Although the sketch is virtually undecipherable to most, Jackson clearly understood what each mark meant, and acted accordingly.

CHAPTER FOUR: FROM FREDERICKSBURG TO CHANCELLORSVILLE

When General Lee crossed the Potomac into Virginia in September 1862, it was without fear of being pursued by McClellan. He rested the army in the valley, healed the wounded, collected fresh supplies, and replaced lost weapons. Stuart's cavalry kept a watchful eye on McClellan, who showed no inclination to stray from his encampment despite being regularly prodded by Lincoln to fight.

Jackson rested his tattered, barefoot corps near Winchester and kept Hotchkiss busy drawing after-action maps from a wealth of sketches covering the long campaign that began at Cedar Mountain and ended at Antietam. General Lee also drew on Hotchkiss's talents, remembering him quite well from the Cheat Mountain campaign in September 1861. The maps produced by Hotchkiss after Cedar Mountain impressed Lee even more, especially the choice of routes around Pope's army that led to the Confederate victory at Second Manassas. Although Hotchkiss remained Jackson's man, Lee frequently borrowed him for special assignments.

HOTCHKISS AND THE CONFEDERATE ARMY

Since joining Jackson during the early stages of the Shenandoah Valley campaign, Hotchkiss had performed his tasks on the general's staff as a civilian noncombatant, without rank and without military status, but with the pay of a first lieutenant. General Jackson addressed him only as "Mister Hotchkiss." Although Hotchkiss served with Lieutenant Boswell as an engineer, he never officially belonged to the Engineer Corps. When Jackson commanded the Valley Army, he could hire whomever he wanted. Once Jackson assumed command of the Second Corps in the Army of Northern Virginia, he lost the privileges of an independent commander and told Hotchkiss he could not keep the topographer on staff without the approval of General Lee. Hotchkiss then asked for a commission, which Jackson did not have the authority to give. With almost continuous fighting during the summer, there had been no time to pursue the matter and Hotchkiss continued to work for Jackson as a civilian. He also continued to collect a lieutenant's pay of $100 per month.

After Antietam, Hotchkiss applied for a commission in the Corps of Topographical Engineers. On September 29 he obtained a letter of recommendation from Lieutenant Boswell, Jackson's staff engineer, and another from the general. "I had a plain talk with the General," Hotchkiss wrote. "[H]e said he thought my great fault was talking too much, but he gave me a good testimonial and hoped that I might succeed in [obtaining] the appointment." After the impromptu performance review, Hotchkiss told Jackson he wished to be retained on the Second Corps' staff, and after the discussion concluded, he packed his letters of recommendation and forwarded them to the Confederate war department.

When more than a month passed without a response from Richmond—although his pay as a lieutenant continued—Hotchkiss hitched his horse to his wagon and on November 4 headed for Lee's headquarters at Culpepper. The Army of the Potomac had finally moved from Maryland and skirmishing once again resumed in the valley. Hotchkiss encountered difficulty locating Lee's headquarters and spent time mapping as he traveled. On November 6 he finally located Lee, who approved Hotchkiss's request to continue as Jackson's topographer. Instead of returning to the valley, Hotchkiss revisited the Cedar Mountain battlefield to survey the area completely so a more thorough map could be submitted with Jackson's battle report. (See page 53, Map #85-4.)

Several weeks passed before Hotchkiss learned that he had been temporarily promoted by the secretary of war to the equivalent of captain of engineers, which provided him with extra pay and a temporary military rank that never became official. He was unaware Jackson had sent a separate letter to the war department extolling his services as a mapmaker and recommending that he be commissioned as a captain. Hotchkiss's official captain's commission never made it through Confederate red tape.

ARMIES ON THE MOVE

Hotchkiss began to work his way back to the valley and sought shelter from a sleet storm mixed with snow. "Soldiers were passing all day," he wrote on November 7, "coming after apple brandy at $15 a gallon; some of them were even barefooted." He crossed through the Blue Ridge three days later and began searching for Jackson's headquarters. After locating the general near Winchester, he enjoyed a long talk. Jackson told Hotchkiss to continue working on maps of the valley and intimated that during the winter the latter might be able to find an office near his Loch Willow home and work there.

Hotchkiss also received a briefing on the movements of the enemy. Much had changed. On November 7 President Lincoln made a decision that stemmed from months of frustration. He removed McClellan as commander of the Army of the Potomac and replaced him with Major General Ambrose Burnside. Because Burnside had previously performed well as an independent commander of a small force, Lincoln believed the general could perform as well leading the Army of the Potomac. Burnside did not want the responsibility because he felt unfit to command so large an army. That McClellan had a low opinion of Burnside's capabilities and blamed him for lost opportunities at Antietam made no impact on Lincoln's decision. All the other generals in the Army of the Potomac were unanimous in their displeasure at the appointment, and internal carping began almost immediately. Although Lee's Army of Northern Virginia had always been confident of beating McClellan, they celebrated the appointment of Burnside, whom they held in even less esteem.

Lincoln wanted action, and he expected Burnside to be aggressive. McClellan had advanced the army at a snail's pace, and when Burnside assumed command, the Army of the Potomac had taken a position where it could interpose between Longstreet at Culpepper and Jackson in the valley. McClellan had planned to defeat one wing of Lee's army and then defeat the other. The strategy made excellent sense, but Lincoln doubted whether McClellan had the fortitude to act. The president expected Burnside to move rapidly, crush the Confederates, and end the war. Instead of following McClellan's plan, Burnside decided to shift his forces east, maneuver through Fredericksburg, cross the Rappahannock River, and capture Richmond. Lincoln clearly stated that Lee's army was the target and not Richmond, but he reluctantly acceded to Burnside's plan because he wanted action. Burnside moved with uncharacteristic speed and sent General Sumner's newly formed Right Grand Division to Fredericksburg. He then moved Franklin's Left Grand Division east of Sumner and repositioned Hooker's Center Grand Division to Falmouth, west of Fredericksburg. By November 19, Federal troops blanketed Stafford Heights from Falmouth to the far side of Fredericksburg.

Lee immediately reacted. He shifted Longstreet's First Corps from Culpepper to Marye's Heights, which overlooked a broad field behind Fredericksburg. Jackson also received a summons. He put his corps on the road, passed through Front Royal, marched through Fishers Gap in the Blue Ridge, and fanned out on roads leading to Fredericksburg. On November 28, after the movement began, Jackson sent orders to his mapmaker to report at Orange Court House. Hotchkiss had fallen ill and went home. He rejoined Jackson on December 4 and found Lee's army spread along broken heights that paralleled the Rappahannock River. Jackson put Hotchkiss to work making a map of Caroline County, which lay between Fredericksburg and Richmond. If for some reason Burnside pushed the Confederates off the hills, forcing their withdrawal, Jackson wanted Hotchkiss to be familiar with the roads leading to Richmond.

FREDERICKSBURG: THE CONTEST BEGINS

On December 11 Hotchkiss wrote, "The artillery firing in the morning woke us up; the cannonading was very heavy . . . The Yankees opened on Fredericksburg . . . just before dawn, and attempted to throw pontoons across the Rappahannock . . . but were driven back." Hotchkiss watched the artillery exchange from a hill and then returned to his quarters to continue working on Jackson's map of routes through Caroline County. The fighting that day occurred at Fredericksburg, where Federals began laying pontoons across the 400-foot-wide Rappahannock. When the morning fog lifted, Brigadier General William Barksdale's Mississippi brigade of sharpshooters, firing from basements and windows, opened on the pontoniers and drove them off. In an effort to repel the sharpshooters, Federal artillery shelled the town and set buildings on fire. With artillery cover and the town in flames, Federal infantry paddled across the river in pontoon boats and eventually cleared the town of sharpshooters.

Jackson expected to be attacked on the morning of December 12 and ordered his army to breakfast at 4:00 a.m. He located Hotchkiss and Howell Brown and took them to Telegraph Road for a quick assessment of the terrain. After watching enemy movements across the Rappahannock, Jackson crossed Prospect Hill and placed A. P. Hill's division on the right of Deep Run and Hood's division from Longstreet's corps on the left. Stuart's cavalry occupied the far right flank east of the tracks of the Fredericksburg and Potomac Railroad. Taliaferro's Stonewall Division occupied the gap between Hill's and Hood's divisions; Early's division occupied the gap between Hill's right flank and Stuart; and Harvey Hill's division occupied the reserve position behind Taliaferro and Early at Hamilton's Crossing.

AMBROSE BURNSIDE (1824–81)

Born in Indiana, Ambrose Burnside obtained an appointment to West Point and in 1847 graduated eighteenth in his class. He served briefly in Mexico and on the frontier before resigning in 1853 to manufacture a carbine he had invented. After his factory in Bristol, Rhode Island, failed, Burnside went to work for George B. McClellan, chief engineer for the Illinois Central Railroad. The friendship with McClellan lasted until the battle of Antietam.

On a social level, "Burn," as he was known, made friends easily because he appeared to be knowledgeable, intelligent, and affable. The magnificent whiskers flowing down his jowls gave rise to the style known as sideburns. He also played poker, often recklessly, and sometimes lost heavily. After rejoining the Union army and commanding a brigade at First Manassas, Burnside performed well during small coastal campaigns in North Carolina. Having established a record as an independent commander, Burnside twice rejected offers from the president to command the Army of the Potomac. The general understood his limitations better than the president.

Placed in command of the Army of the Potomac's left wing at Antietam, Burnside demonstrated indecisiveness. He failed to push his men in a timely manner across a stone bridge that probably would have given McClellan a significant victory. The general's friendship with McClellan ended at "Burnside Bridge" and the two seldom spoke afterward. When notified that Burnside had replaced him, McClellan lamented, "They have made a great mistake. Alas for my poor country!"

Burnside had enough trouble commanding a corps without assuming the responsibility for commanding an army. He wanted to refuse the position for a third time, but his friends urged him to take the post. Within days, Burnside could neither sleep nor think strategically. When pontoons promised by the secretary of war failed to arrive on time, Burnside lost the element of surprise and began to panic. Instead of forcing General Sumner to cross the Rappahannock before Jackson's corps arrived from the Shenandoah, Burnside choked. Through stress and fatigue, he became unable to issue crisp and timely orders in language his generals understood. Although Burnside held other commands after the battle of Fredericksburg, all of his weaknesses as a commander became manifest when he attacked the Army of Northern Virginia on December 13, 1862.

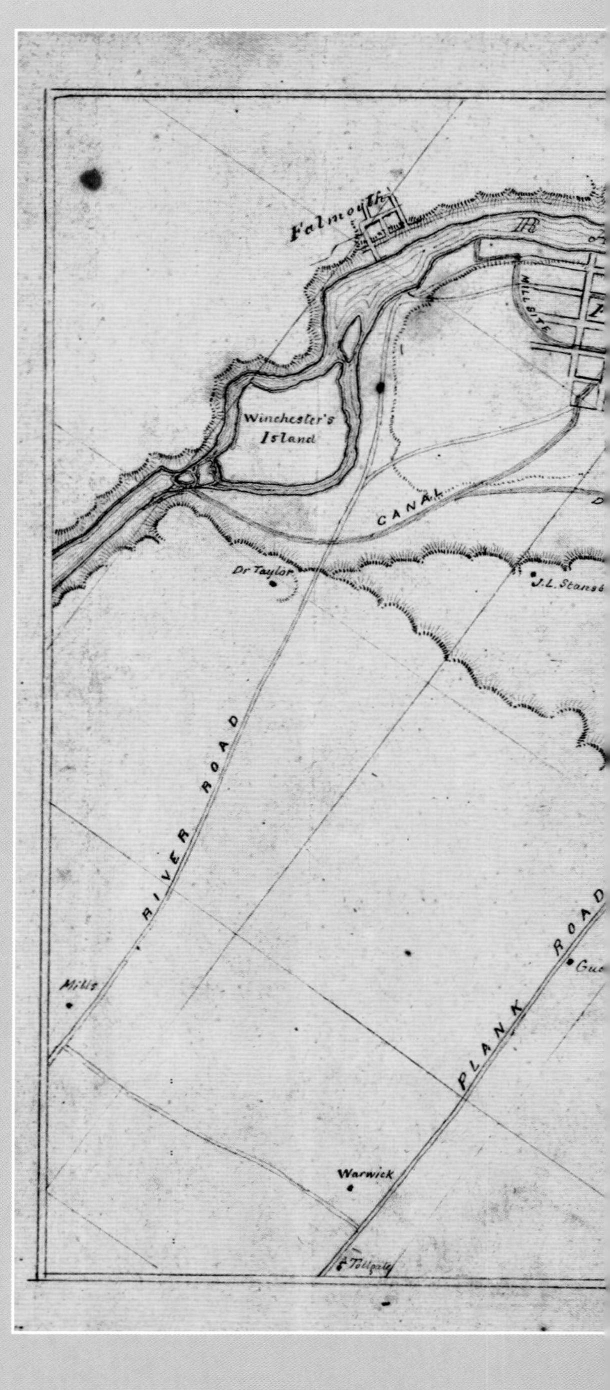

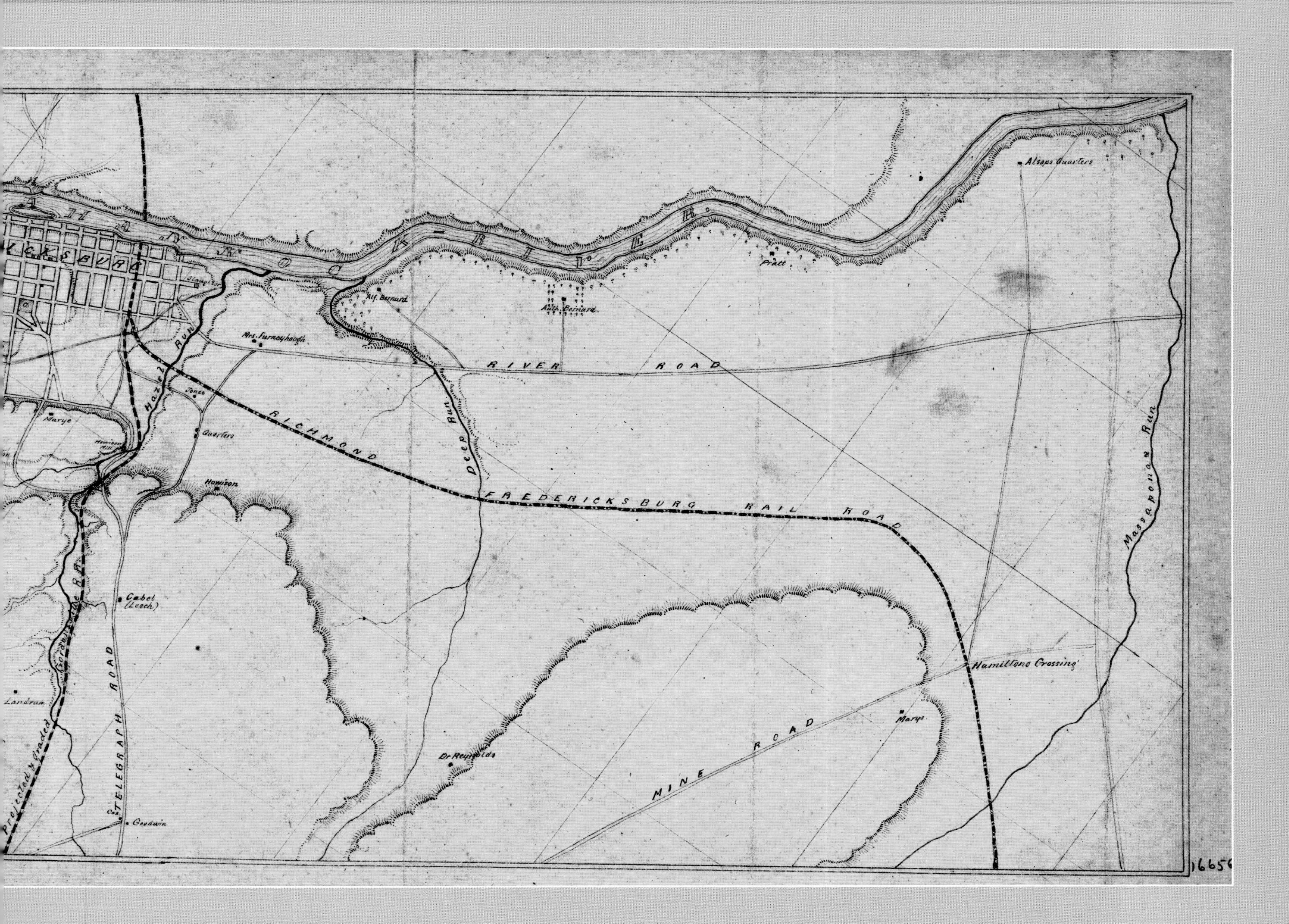

Battle of Fredericksburg: Preliminary Sketch [Map #124]

Before the battle of Fredericksburg began, and before either side deployed their forces, Hotchkiss drafted a quick sketch of the ground that on December 13, 1862, became one of the bloodiest battlefields of the war. Hotchkiss made the sketch to show the roads, streams, and farms in the alluvial plain where the battle occurred. Fredericksburg became a strategic town because the Telegraph Road and the Richmond and Fredericksburg Railroad led directly to the Confederate capital.

Longstreet's First Corps occupied Marye's Heights with Richard Anderson's division on the left followed sequentially to the right by the divisions of McLaws, Robert Ransom, and George E. Pickett. Lee placed Pickett's division between Hood and Ransom, making the unit available to support either Jackson or Longstreet. Considering that Lee's 78,500-man army stretched over seven miles, one would have expected the Federals to have probed for gaps. The only fighting that day was in Fredericksburg, where Federal forces finally ousted the last of Barksdale's sharpshooters and took possession of the town.

That evening, Burnside's 122,000-man Federal force occupied the far side of the Rappahannock from Falmouth to Deep Run. The First and Sixth Corps of Franklin's Left Grand Division had thrown three pontoon bridges across the river near Deep Run. At Fredericksburg, three more pontoon bridges stretched across the river into Fredericksburg for conveying Sumner's Right Grand Division and Hooker's Center Grand Division into Fredericksburg. As night fell, all three grand division commanders still waited for orders from Burnside. Behind the infantry and atop the hills, Brigadier General Henry J. Hunt's batteries, consisting of 147 guns, lay arrayed across Stafford Heights with most of the muzzles bearing on Marye's Heights.

The Federals spent a restless night waiting for Burnside to issue orders for the assault.

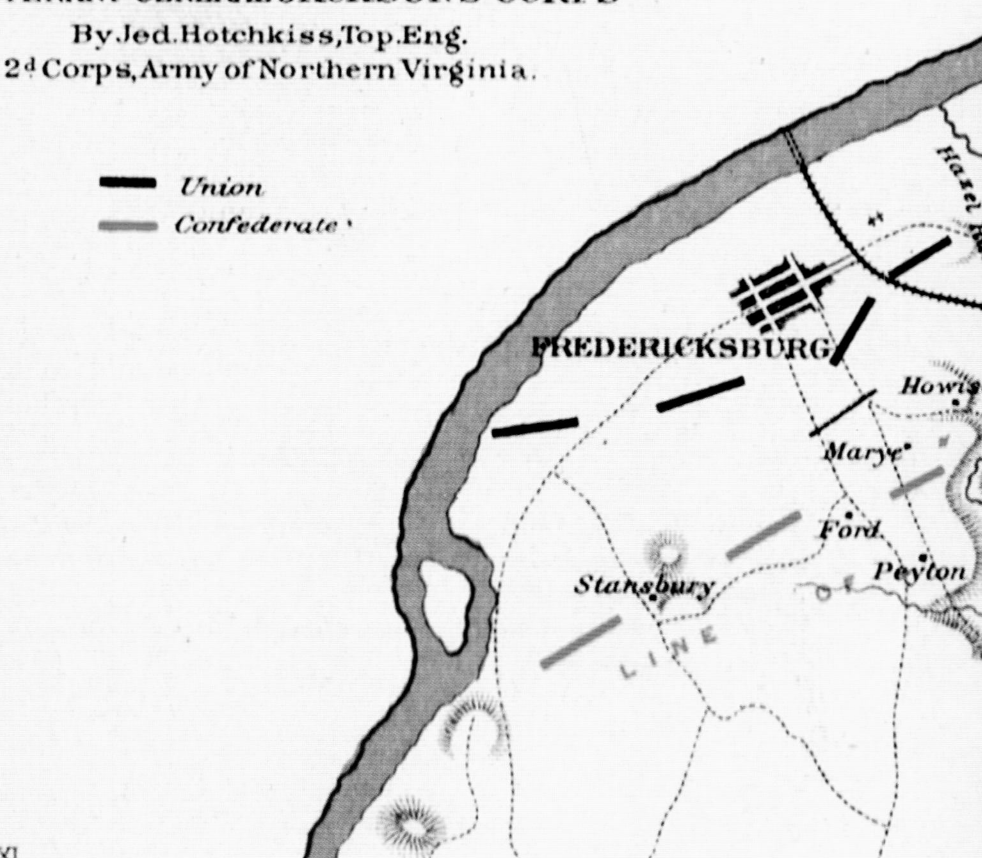

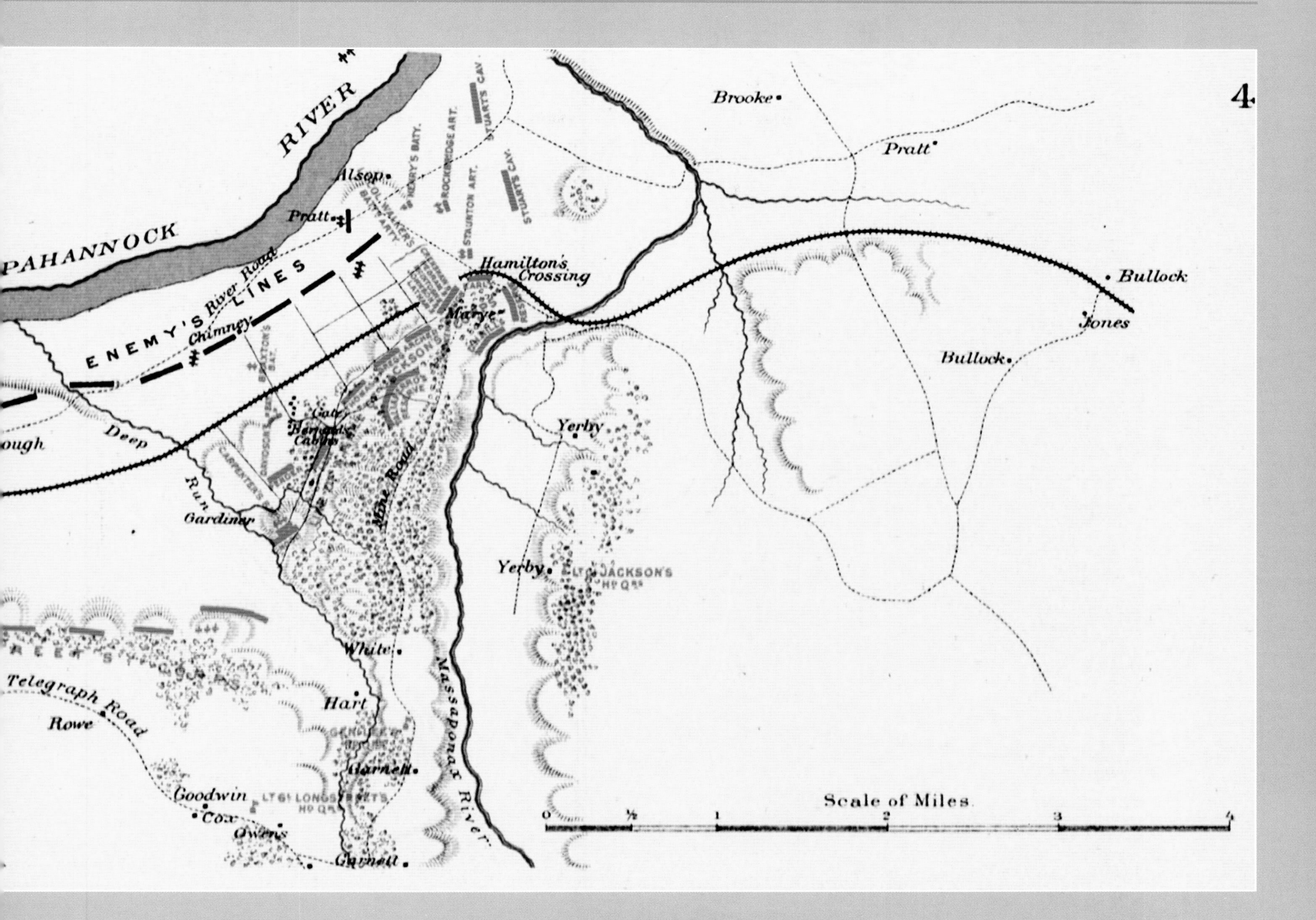

Battle of Fredericksburg [Map #31-4]

Hotchkiss made sketches of the battlefield but never completed a map of the action until General Jackson asked that one be made for his battle report. The map Hotchkiss eventually made for the Official Records *has served over the years as the template used by hundreds of authors of articles and histories chronicling Jackson's fight during the battle of Fredericksburg.*

THE ASSAULT ON THE CONFEDERATE RIGHT

According to Hotchkiss, Jackson had been spoiling for a fight ever since Burnside began throwing pontoons across the Rappahannock. When Hotchkiss rode with the general to survey the field on December 12, they went almost as far as the river without encountering picket fire. Through lifting fog, the general saw thousands of Franklin's troops filing across pontoons. He expected the first blow to come on the right flank and was so pleased with the Federal dispositions that he "whistled as we went along."

"We were up at an early hour and off to the battle field by daylight," wrote Hotchkiss. "[T]he tents were struck and the wagons loaded up and sent to the rear. All our troops were in position early." Hotchkiss did not mention that Jackson's corps then waited six hours for the Federals to attack.

The delay on December 12 occurred because Burnside called for a general assault at dawn without issuing clearly comprehensible orders. The delay on December 13 occurred because of heavy fog in the morning and more confusion over what Burnside meant by his orders. In essence, Burnside's plan called for Sumner's two corps and Hooker's two corps to assault Longstreet at the same time Franklin's two corps attacked Jackson. Sumner, Hooker, and Franklin still had questions about their orders when, at 10:00 a.m., the fog burned off, sun flooded the battlefield, the temperature quickly rose, and the mighty panorama of the battlefield spread before the eyes of Confederates waiting in the hills.

Around noon, Federal artillery opened and shells began bursting among Jackson's corps waiting on Prospect Hill. Thirty minutes later, with flags fluttering and polished rifles brightly shimmering in the sunlight, 50,000 troops from Franklin's force formed for an attack on A. P. Hill's brigades positioned behind the embankment of the Fredericksburg and Potomac Railroad. Franklin pushed forward only John Gibbon's and George Meade's divisions, which forcibly struck Maxcy Gregg's brigade from Hill's division. Dense woods caused Gibbon to become separated from Meade, who accidentally strayed into a gap several hundred yards wide and unprotected by Southern troops. Meade began funneling troops through the breach. He called for reinforcements, and made every effort to keep the corridor open until supports arrived from Franklin. Meade held on as long as possible, but Daniel E. Sickles's and David B. Birney's divisions did not move rapidly enough to take advantage of the gap. Gibbon's assault had moved several hundred yards to the right of Meade and became blocked by terrain. Jackson acted quickly. Four Confederate brigades from Early's and Taliaferro's divisions counterattacked, met the Federals head-on, and, aided by Stuart's horse artillery on the far right, sent the Federals reeling out of the woods and sealed the breach. After the grayclads decimated two late-arriving reserve regiments from Birney's division, the Federals retired back across the railroad tracks.

For unaccountable reasons, Franklin did not press the attack and seemed content to hold his position rather than assault Jackson. Whether Franklin's disdain of Burnside and his former close relationship with McClellan blunted his enthusiasm for fighting might never be known. Franklin followed his commander's hazy orders literally and failed to dislodge Jackson. When Burnside needed someone to blame for the fiasco at Fredericksburg, he pointed squarely at Franklin.

Two hours after Franklin launched the Federal assault, both combatants were back where they started. After driving the Federals across the railroad embankment, Jackson considered attacking Franklin and pushing the Federals back across the river. He fortunately withheld the temptation because of Longstreet's heavy engagement on the left. Most of Franklin's divisions had not fired a shot and would likely have severely damaged Jackson's corps.

THE ASSAULT ON THE CONFEDERATE LEFT

On the left side of the Confederate line, Longstreet occupied a strong defensive position on Marye's Heights. Between the base of the hill and Fredericksburg lay a broad, open plateau. A sunken road, running between Telegraph Road and Orange Plank Road, curved along the base of Marye's Heights. On the Fredericksburg side of the road ran a shoulder-high stone wall banked with earth, which formed the most impregnable defensive position on the entire field. A soldier firing from behind the stone wall did not have to expose more than his head and shoulder when firing his weapon. Longstreet clearly understood the advantage of the position. He lined the sunken road with Brigadier General Thomas Cobb's 2,000-man Georgia brigade from McLaws's division and placed Brigadier General Joseph B. Kershaw's brigade in reserve behind the ridge. Cobb positioned his Georgians in two ranks. While one man fired, the other drew back from the wall and reloaded.

On the heights above the sunken road, Longstreet placed artillery to cover the battlefield. His defensive deployments made it almost impossible for Federal batteries to inflict much damage. Longstreet also covered his flanks. On the right, where the stone wall ended, he prepared a trench for Brigadier

General Robert Ransom's Twenty-fourth North Carolina. If the enemy's field commanders hoped to breach the Confederate defenses at the weakest point, they would first have to cross the stone wall held by Cobb.

About noon, Longstreet opened fire on the streets of Fredericksburg as a diversion to assist Jackson, whose forces were under attack on the right. The shelling stirred up a hornet's nest. General Sumner gave the order to advance, and blueclads from Nathan Kimball's brigade, which had been concealed behind homes, "swarmed out of the city like bees out of a hive," said Longstreet. Kimball's brigade charged Cobbin in double-quick formation. "The field was literally packed with Federals," Longstreet recalled. "From the moment of their appearance began the most fearful carnage. With our artillery from the front, right, and left tearing through their ranks, the Federals pressed forward with almost invincible determination, maintaining their steady step and closing up their broken ranks."

Major General William H. French, leading the Third Division of Major General Darius Couch's Second Corps, formed the division in the line of battle before advancing on the sunken road and ordered his blueclads to fix bayonets. One Confederate artilleryman watching from the heights commented, "How beautifully they came on. Their bright bayonets glistening in the sunlight made the line look like a huge serpent of blue and steel. We could see our shells bursting in their ranks, making huge gaps. But on they came, as though they would go straight through and over us."

As Kimball's brigade closed on the stone wall, Cobb's infantry opened in a sudden sheet of flame at 125 yards. Another Confederate volley followed, and then another. A few Federals came within forty yards of Cobb's position, firing and reloading before falling or running for cover. In less than twenty minutes, a quarter of Kimball's brigade had been killed or wounded. Kimball had to be carried off the field with a bullet in his thigh.

Despite the horrendous losses in Kimball's brigade, French continued to send his brigades into a bloodbath of Confederate fire. Colonel John W. Andrew's Third Brigade lost half its men in fifteen minutes. Colonel Oliver H. Palmer's brigade came no closer to the sunken road and suffered similar losses. By then, French's division had been squandered and shot to pieces. But Burnside was not done. Couch sent in Hancock's division, and when Hancock's men failed to breach the stone wall and piled up on the battlefield, Sumner and Hooker lined up three more infantry divisions for the slaughter. Hancock's division, one of the best in the Army of the Potomac, lost 2,049 men in an hour, roughly 42 percent of its strength. The common soldier quickly realized the hopelessness of the charges and avoided the carnage by hugging the ground behind so little as a clump of grass.

"The charges had been desperate and bloody," said Longstreet, "but utterly hopeless." He counted at least six charges. A few Federals almost reached the stone fence before being shot. Cobb fell midway through the afternoon, and McLaws sent in Kershaw's brigade. As dusk descended late that afternoon, the field in front of the sunken road lay covered with dead piled one on top of the other, with some three deep. Fresh Federal regiments pouring onto the field simply dropped behind the dead and dying to protect themselves from Confederate fire.

With nightfall came the cold and the agonizing groans and whimpers of thousands of Federal wounded. When morning broke on December 14, the frosted ground was still covered with the dead, countless having died during the night from wounds and exposure. Burnside considered renewing the attack on Marye's Heights in the morning, but General Hooker snorted in disgust. "There has been enough blood shed to satisfy any reasonable man, and it is time to quit," said Hooker. On the night of December 15, the Federals evacuated Fredericksburg and crossed to the far bank of the Rappahannock. As the casualty reports arrived at Union headquarters, the carnage at Fredericksburg staggered Burnside.

BATTLE OF FREDERICKSBURG CASUALTIES

	Engaged	Killed	Wounded	Missing	Total
Confederate	72,500	608	4,116	653	5,377
Federal	106,000	1,284	9,600	1,769	12,653

THOMAS READE ROOTES COBB (1823–62)

COBB NEVER RECEIVED ANY FORMAL MILITARY TRAINING AND SPENT HIS PROFESSIONAL CAREER BEFORE THE WAR AS A TALENTED AND BRILLIANT CONSTITUTIONAL LAWYER IN GEORGIA. HE WROTE NUMEROUS BOOKS ON LAW, PUBLISHED A DIGEST OF THE LAWS OF GEORGIA, AND CONTRIBUTED MANY ARTICLES TO NEWSPAPERS AND MAGAZINES. COBB ALSO BECAME ONE OF THE OUTSTANDING VOICES, WITH HIS BROTHER HOWELL COBB, FOR TAKING GEORGIA OUT OF THE UNION.

WHEN THE CIVIL WAR BEGAN, COBB SURPRISED MANY OF HIS FRIENDS BY REJECTING POLITICAL OPPORTUNITIES TO FORM A REGIMENT. THOUGH COMPLETELY LACKING IN MILITARY EXPERIENCE, HE BECAME A COLONEL IN CHARGE OF A MIXED COMMAND KNOWN AS COBB'S LEGION. HE LEARNED HOW TO LEAD A REGIMENT BY ON-THE-JOB TRAINING DURING THE PENINSULA CAMPAIGN. AT ANTIETAM, WHERE HIS MEN SUFFERED SEVERELY, HE BEGAN TO UNDERSTAND WARFARE.

APPOINTED BRIGADIER GENERAL ON NOVEMBER 1, 1862, COBB ASSUMED COMMAND OF A BRIGADE IN LAFAYETTE MCLAWS'S DIVISION. SIX WEEKS LATER, GENERAL LONGSTREET TASKED COBB WITH DEFENDING THE STONE WALL IN THE SUNKEN ROAD AT FREDERICKSBURG. HE STUBBORNLY HELD THE POSITION AGAINST A CRUSHING FEDERAL OFFENSIVE THAT TRIED SIX TIMES TO ASSAULT AND OVERWHELM HIS 2,000-MAN BRIGADE. DURING ONE OF THE ASSAULTS, A BULLET SMASHED INTO COBB'S THIGH. HE WAS TAKEN TO A HOUSE ON THE BATTLEFIELD FOR MEDICAL AID, BUT THE WOUND BLED OUT AND COBB DIED. PROBABLY NO MAN WHO COMMANDED A BRIGADE DURING THE WAR EVER FOUGHT AGAINST SO MANY FEDERALS AT ONE TIME WITHOUT ABANDONING HIS POSITION.

WINTER MAPPING

The December weather turned suddenly cold, and the Confederates began building winter quarters. Hotchkiss spent a few days wandering the battlefield and on occasion observed a dead blueclad lying in a crevice and covered with light snow. He felt remorse at the sight of his own countrymen, though they came as enemies from the North, lying forgotten on a barren battlefield, a picnic for crows and turkey vultures. He felt less remorse toward the enemy when he rode into Fredericksburg and found parts of the city leveled by Federal artillery and townspeople wandering the streets in despair.

Having seen enough of the evils of war, Hotchkiss pitched his tent near Jackson's headquarters on the grounds of Richard Corbin's palatial home at Moss Neck and returned to his maps. He still needed information to complete the map of Caroline County and sent two of his assistants, Williamson and Kinney, to survey routes, especially those that led to Richmond. He discovered many inaccuracies in an old Caroline County map used for reference.

Jackson had forwarded a considerable amount of correspondence to the war department regarding operations over the past year, but he had yet to file an official report on his campaigns. Having grown weary of reminders from the war department for reports, Lee attached Colonel Charles A. Faulkner as an aide to Jackson in collecting the documents. As winter was the best time to get this onerous task out of the way, Jackson put Hotchkiss to work converting the sketches into dignified after-action battle maps.

Hotchkiss's journal became a litany of the general's requests. On January 7, 1863, Hotchkiss wrote: "Worked on the Battle of Kernstown most of the day." Two days later: "Re'cd an order . . . for a map of the Battlefield of Fredericksburg." (See page 77.) Hotchkiss lacked battlefield information and spent days collecting the details from brigade and regimental commanders. On January 10 General Stuart appeared at Hotchkiss's tent and asked for a map of the battle of Groveton, by which he meant the battle of Second Manassas (see page 62, Map #111-1) for Jackson. During the war, Stuart developed a deep appreciation for the value of accurate maps. He established a close friendship with Hotchkiss and became a regular customer. Once other commanders discovered Hotchkiss's huge collection of maps and sketches, they began calling at his tent for sketches to insert with their own reports.

Colonel Faulkner, Lee's appointed collector and assembler of reports, added his own requests to Hotchkiss's workload. Although Hotchkiss had never been on the field during McClellan's Peninsula campaign, Faulkner wanted a map of Richmond and vicinity. Hotchkiss hurried through the work using old maps of the capital and, having reached the point of exhaustion, received permission from Jackson for a two-week leave.

On returning to the Fredericksburg area on February 19, Hotchkiss found Faulkner waiting with a stack of reports for the topographer's comments. Some of the generals could not remember details and wanted their battle reports confirmed by Hotchkiss. Faulkner decided that Jackson should have a map of the battle of Winchester. Hotchkiss had a large volume of sketches and produced another map (see page 37, Map #85-2). Faulkner also asked for a map of the battle of Cedar Mountain, which Hotchkiss had also started; a few days later he produced a refined map (see page 53, Map #85-4).

Faulkner finished his report-collecting effort at the end of February and asked to be transferred to other duties. Hotchkiss noted in his journal a feeling of relief when the report writing ended. He planned to give himself a break, but in late February Jackson asked for a map of Cumberland County, Pennsylvania. Why Jackson wanted a map of Cumberland County puzzled Hotchkiss, until he noticed that roads led to Harrisburg, the capital of Pennsylvania. Jackson then requested a map of Fauquier County, Virginia, which contained routes along the east side of the Blue Ridge Mountains that also stretched north. For the next several weeks, Hotchkiss worked relentlessly on maps, mostly of counties through which the army would later pass. Many of those maps would soon become part of another massive effort to update the roads, bridges, and terrain reaching into Maryland and beyond.

HOOKER REPLACES BURNSIDE

First there was "Burnside Bridge" and then came Burnside's "Mud March." The Confederates watched the debacle from the south side of the Rappahannock. In a poorly conceived campaign to redeem himself, Burnside decided to march up the river, cross at Banks's Ford, and strike Lee from the rear. On January 19, 1863, advance units moved out from their camp on Stafford Heights. Almost as if on cue, the skies opened with two days of cold torrential rain and high winds. Near Banks's Ford, the long and disorganized column slogged to a halt, stopped by knee-deep mud. Supply wagons upset, artillery sank to their axles, ammunition trains toppled, and horses and mules collapsed and drowned in the liquid mud. Confederates watching from across the river erected signs with mocking messages such as "Yanks, if you can't place your pontoons yourself, we will send help." On January 26, after receiving the news of Burnside's last blunder, Lincoln replaced the general with Joseph Hooker.

Hooker had performed well as a division or corps commander, but he manifested a bad habit of always criticizing the conduct of his superior. Burnside's incompetence infuriated him, and he liberally broadcast his feelings. When Lincoln decided to change commanders in the Army of the Potomac, he praised Hooker's fighting ability but criticized him

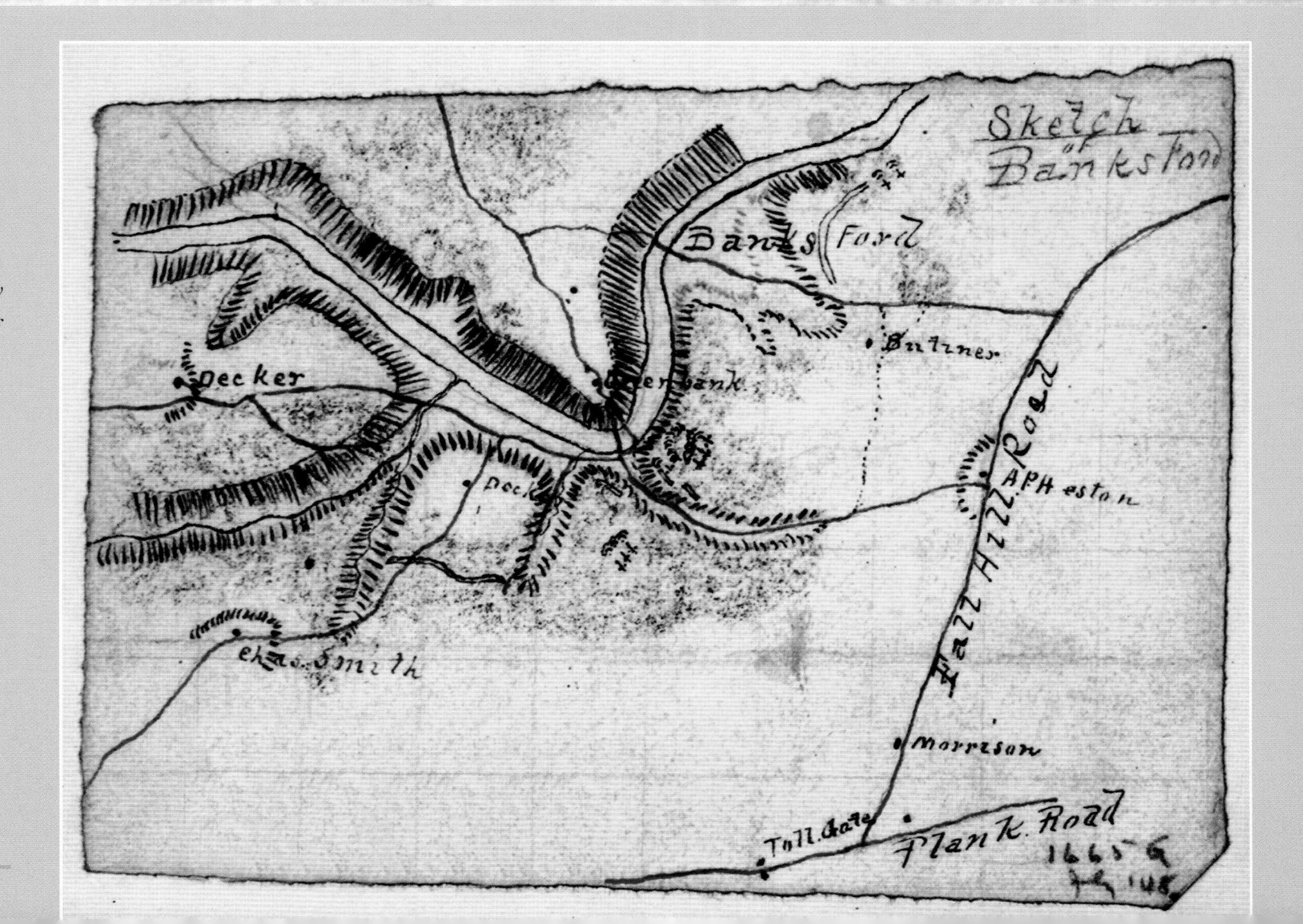

Banks's Ford [Map #139]
Located five miles west of Fredericksburg, Banks's Ford provided a good crossing for both armies when the level of the Rappahannock River was normal. Hotchkiss made many copies from his single sketch of the ford for Stuart's cavalry, which constantly used and patrolled the crossing.

for not cooperating with Burnside. He also instructed Hooker to concentrate on destroying Lee's army instead of trying to capture Richmond.

During the Peninsula campaign, a copyist for the Associated Press ran an article on Hooker's May 5, 1862, assault on Williamsburg. The heading read "Fighting—Joe Hooker." Newspapers throughout the country changed the heading to read "Fighting Joe Hooker," and the nickname stuck. Lee sarcastically referred to Hooker as "F.J.," which Hotchkiss found amusing. He also observed that Lee looked forward to testing F.J.'s fighting skills.

THE NEW ARMY OF THE POTOMAC

After Fredericksburg and the infamous "Mud March," no Union army had suffered more humiliation than Lincoln's vaunted Army of the Potomac. Morale diminished, and some generals resented the elevation of Hooker, whom they felt had capitalized on Burnside's ineptness to get the job. Hooker set his sights even farther down the road, predicting that the winning general might rightfully claim a postwar dictatorship. Lincoln acknowledged Hooker's comment by writing, "What I ask of you is military success, and I will risk the dictatorship."

Hooker rebuilt the Army of the Potomac. He abolished Burnside's grand divisions and organized the cavalry into a corps under Major General George Stoneman. He developed a system of distinctive badges for each corps, restored morale, and increased the strength of the army to 134,000. By the spring of 1863, Hooker bragged that he had "the finest army on the planet." He also developed a brilliant strategy for his spring campaign. He would not make Burnside's error of shoveling in brigades piecemeal. He would use his entire force and envelop Lee's army, part of which still occupied the hills south of Fredericksburg.

HOOKER'S GRAND MANEUVER

Hooker's strategy looked good on paper. General Stoneman's cavalry would splash across an upper ford, circle deep behind the Confederate position, and cut Lee's communications. General Meade's Fifth Corps, General Howard's Eleventh Corps, and General Slocum's Twelfth Corps would march up the river and cross over Kelly's Ford on the Rappahannock. Meade would cross at Ely's Ford on the Rapidan, with Howard and Slocum crossing at Germanna Ford, also on the Rapidan. General Sickles's Third Corps and General Couch's Second Corps would cross the Rappahannock at U.S. Ford; join up with Meade, Howard, and Slocum at Chancellorsville; and with Hooker and two-thirds of the army would countermarch down the Plank Road and the River Road to attack Lee's left flank. General Sedgwick's Sixth Corps and General Reynolds's First Corps, with the other third of the army, would cross the Rappahannock below Fredericksburg and make a diversionary attack against Jackson's front. Hooker's plan had excellent features, providing Lee made no countermoves.

Hooker's finely tuned strategy encountered its first pitfall on April 13 when Stoneman's 10,000-man cavalry rode out of Falmouth but found the Rappahannock fords flooded. The cavalry did not wade the river until April 29, the same day the infantry crossed. Stoneman's force disappeared into Virginia too late to disrupt Lee's communications but just in time to vanish when Hooker most needed the eyes of his cavalry.

During a recent visit from Lincoln, Hooker had bragged that the question was not whether his army would capture Richmond, but how soon. Despite Stoneman's disappearance, on April 29 Federal forces brushed Confederate pickets and skirmishers out of the way and marched by different roads toward Chancellorsville. Two days later, advance units reached the befuddling Wilderness, which was located about twelve miles west of Fredericksburg. Aptly named, the Wilderness was a formidable, thickly wooded, brush-covered region with hills, twisting roads, and logging trails. By daylight on May 1, Hooker had 75,000 blueclads near Chancellorsville and a force of 40,000 at the lower pontoon bridges where Franklin's corps had crossed during the Fredericksburg campaign.

Hooker began to assemble his forces to assault Lee's left flank when Jackson's Second Corps unexpectedly appeared in the vicinity of Newton on the Old Turnpike. Hooker had received no reports of Jackson's movements because he had sent his cavalry corps on a wild-goose chase. The news startled him, so instead of continuing with his plan, he decided to fight the Confederates at Chancellorsville.

U.S. (United States) Ford [Map #140]

Hotchkiss made several sketches of the fords strung along the Rappahannock and Rapidan rivers. Located just below the confluence of the two rivers, U.S. Ford was a long, broad stretch of shallows used for years by local traffic. During the war, it became a much-traveled ford for infantry and cavalry.

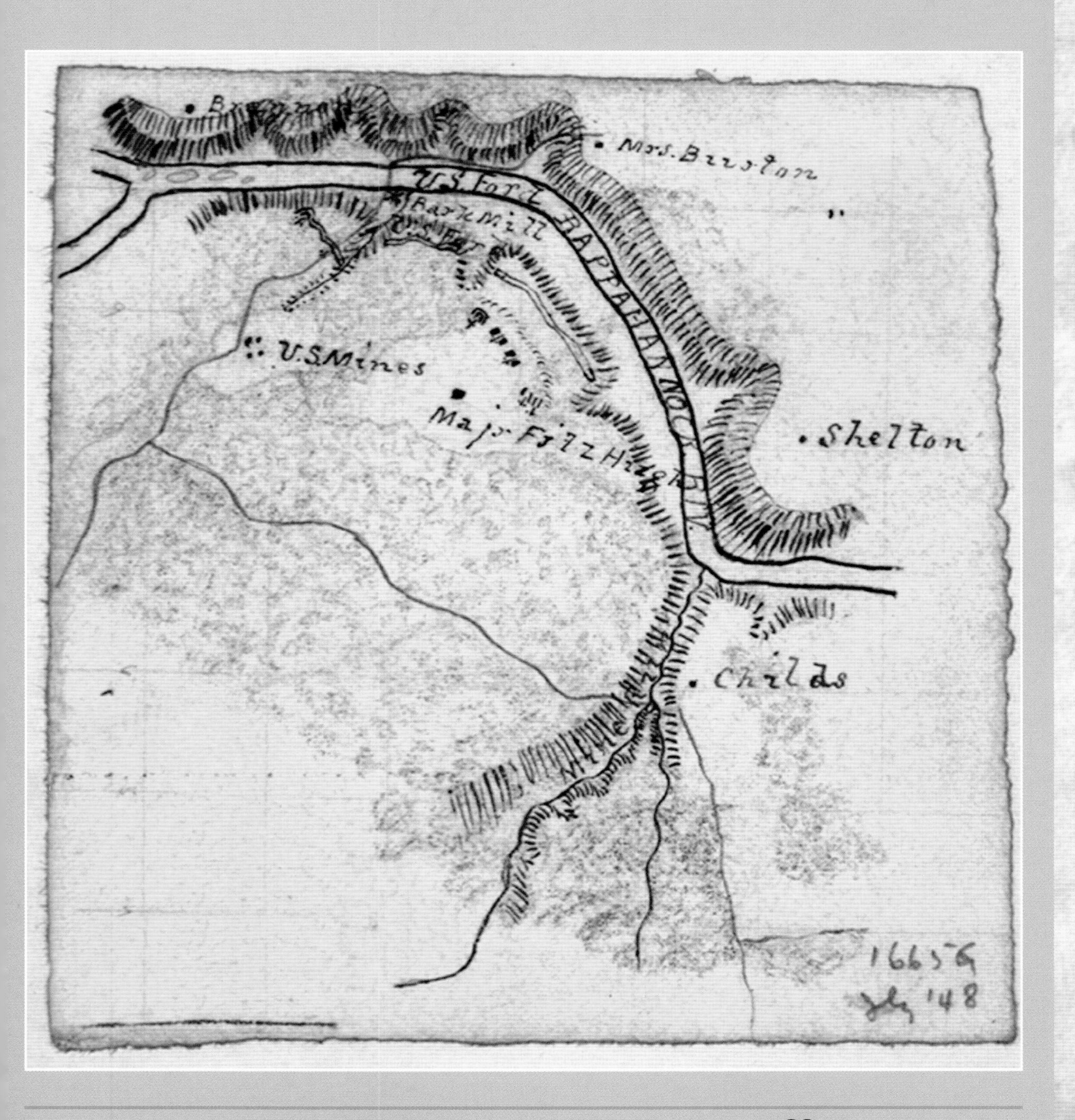

Major General Joseph Hooker (1814–79) *replaced General Ambrose Burnside as commander of the Army of the Potomac after accusing the latter of incompetence. Hooker bragged about capturing Richmond with "the finest army on the planet" a few weeks before Stonewall Jackson crushed the Federal right flank at Chancellorsville and destroyed Hooker's hopes.*

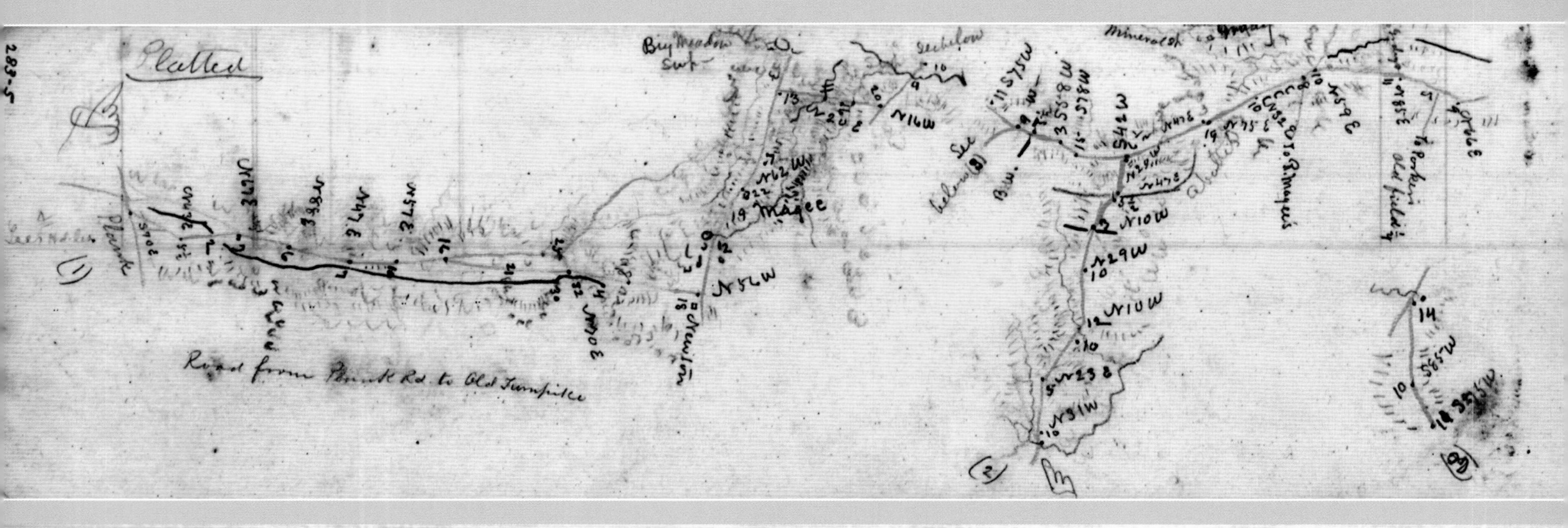

LEE'S COUNTERMANEUVER

Early on April 29, Hotchkiss heard firing along the Rappahannock. As the fog lifted at noon, he observed Federals crossing and extending their line on Jackson's right flank. Jackson moved Early's division into a blocking position and waited for developments. Later in the day, scouts reported masses of Federals crossing at the upper fords of the Rapidan.

Lee had a distinct disadvantage. On February 17, 1863, Longstreet had been sent with Pickett's and Hood's divisions to the Suffolk, Virginia, area to command the Confederate Department of North Carolina and Southern Virginia. Longstreet showed little aggressiveness and no aptitude for independent command. The other half of his corps, Anderson's and McLaws's divisions, remained with Lee.

The following day, Jackson asked Hotchkiss to "strike off eight maps embracing the region between the [Rapidan] and the Rappahannock and reaching back to the Virginia Central Railroad." Hotchkiss finished the assignment and had just settled for the evening when Jackson reappeared at 9:00 and asked him to scout routes to Salem Church, which was about halfway to Chancellorsville. When Hotchkiss returned on the morning of May 1, he found everyone preparing to march. Boswell picked up a set of sketches from Hotchkiss's night of scouting and guided the army down the Plank Road until encountering resistance about three miles from Chancellorsville.

Early on the morning of May 2, the generals held a conference while sitting on cracker boxes in the pines at the top of a hill overlooking Catherine Furnace. Hotchkiss had sketched part of the area, but not all. Before daylight, he went to General Stuart's headquarters to obtain a cavalry escort. Knowing Jackson's penchant for launching flank attacks, Hotchkiss searched through several back roads and found a route that led to the enemy's rear. He also spoke to a local blacksmith, Charles B. Wellford, who mentioned that he had recently cut new roads through the forest for hauling wood and

Lee's Countermaneuver [Map #142-8]
Hotchkiss made numerous sketches of the roads (in red) through the Wilderness, and many of them today cannot relate to a specific Wilderness area. This sketch is more decipherable because the lower half begins at Newton, which is on the Old Turnpike about 1.5 miles east of Chancellorsville, and the road leads south to the Plank Road, which is where Jackson began his flank movement (black line). The top map has only one reference to the lower map, that being a farm owned by Magee, but because there were two Magee farms on opposite sides of the turnpike, the exact location of these roads is not clear. Another point of interest is Hotchkiss's tall, slender sketchbook. He always needed paper, and this sketchbook is an example of simply using whatever paper he had available.

iron to his furnace. Hotchkiss returned from his reconnaissance, placed a cracker box between the two generals, and laid out his sketch. Hotchkiss explained the sketch by using it together with a map that he had previously given to Jackson. Lee listened and raised his head. "General Jackson," he asked, "what do you propose to do?"

Jackson pointed to a route on Hotchkiss's map and replied, "Go around here."

"What do you propose to make this movement with?" Lee asked.

"With my whole corps," Jackson replied.

Lee then asked what troops Jackson intended to leave behind, and the general replied, "The divisions of Anderson and McLaws."

Despite the danger of dividing his force, Lee replied, "Well, go on."

Jackson formed the Second Corps with Rodes's division leading off and with Hotchkiss as a guide. He asked for quiet, and as the long column got underway, the only sound heard was the rhythmic clanking of canteens and military impedimenta as the men passed in file. Lee recalled Hotchkiss to field headquarters to request more maps. When finished with the assignment, Hotchkiss used the same route, riding hard to catch up with Jackson. "[We] went on to the Brock road," wrote Hotchkiss, "and then up it a piece and into a private road, and so on to the Plank Road and across both Plank Roads to the Old Turnpike and formed our line of battle at the [Lucket's] house [near Wilderness Tavern] and with three lines of battle fell on the enemy's rear." When he had found Jackson near the Old Turnpike, the general made a request for him to make sketches as the assault progressed and send them back to Lee by courier.

To execute this maneuver, Lee abandoned all the tenets of warfare by splitting his smaller force, already confronted by a superior force, a second time by turning Jackson loose on a march that would consume the entire day.

THE CHANCELLORSVILLE CAMPAIGN: MAY 2, 1863

On the morning of May 2, 1863, 26,000 men from Jackson's Second Corps marched in a formation ten miles long over a semicircular, fourteen-mile concealed route that began near Newton on the Old Turnpike. Lee posted another 14,000 troops along the Furnace Road to check Hooker's advance. General Early's division still occupied the heights near Fredericksburg, slowing the advance of two Federal corps commanded by Sedgwick and Reynolds. Hooker's withdrawal into the woods had baffled Lee. At first he could not decide whether Hooker had been luring the Army of Northern Virginia into a trap or had simply become timid.

Early that evening as the sun began to descend, Jackson's column re-formed near Wilderness Tavern to strike Hooker's right flank. Because General Stoneman's 10,000-man cavalry had disappeared on a belated raid to disrupt Lee's communications, Hooker had blinded himself by retaining too few cavalry to properly scout the army's right flank. General Howard, whose Eleventh Corps guarded the flank, did have pickets, skirmishers, two regiments of infantry, and the reserve artillery stretched along the Old Turnpike west of Chancellorsville, but most of his forces looked eastward, the direction from which he expected to confront the enemy. Over the objections of his corps commanders, Hooker had withdrawn his forces into the woods, neutralized his superiority in numbers and artillery, turned the initiative over to Lee, and later would admit he had lost his nerve.

Federal scouts working through the back roads of the Wilderness did observe Confederates moving on the Brock Road and the Orange Plank Road and reported it to Hooker, who in turn relayed the information to Howard. Howard dismissed the intelligence because Confederate cavalry had been active in the area. He did not know that Stuart's cavalry were on every road screening Jackson's march. Union cavalry operating on the east side of the Wilderness also observed a Confederate column with ambulances and wagons marching westward. Hooker jumped to the conclusion, perhaps through wishful thinking, that Lee was retreating toward the Shenandoah Valley and notified Howard, who accepted the general's suggestion as probable and left his flank hanging in the air.

Shortly after 5:00 p.m., while Howard's blueclads lit fires and cooked their meals, the first indication of trouble occurred when frightened deer, foxes, and rabbits began running through Federal camps. Seconds later, picket fire shattered the late-afternoon quiet, bugles blew, the rebel yell shrieked through the woods, and Rodes's brigade burst upon the 153rd Pennsylvania and the Fifty-fourth New York. The Federals fired three volleys, temporarily slowing the Confederate advance, and fled. The rest of Harvey Hill's brigades stormed through the woods and struck the Forty-first and Forty-fifth New York regiments. Both regiments broke to the rear without firing a shot.

Of the many Federal regiments posted along the Old Turnpike, only Colonel Robert Reiley of the Seventy-fifth Ohio had considered the possibility of a Confederate flank attack. When the grayclads came within thirty paces, the Seventy-fifth Ohio stood in formation and opened with devastating fire. Confederates flanked the position and killed Reiley and 150 blueclads. As evening approached, the Confederates gained momentum, pushing through Federal brigades and driving the enemy back on Wilderness Church.

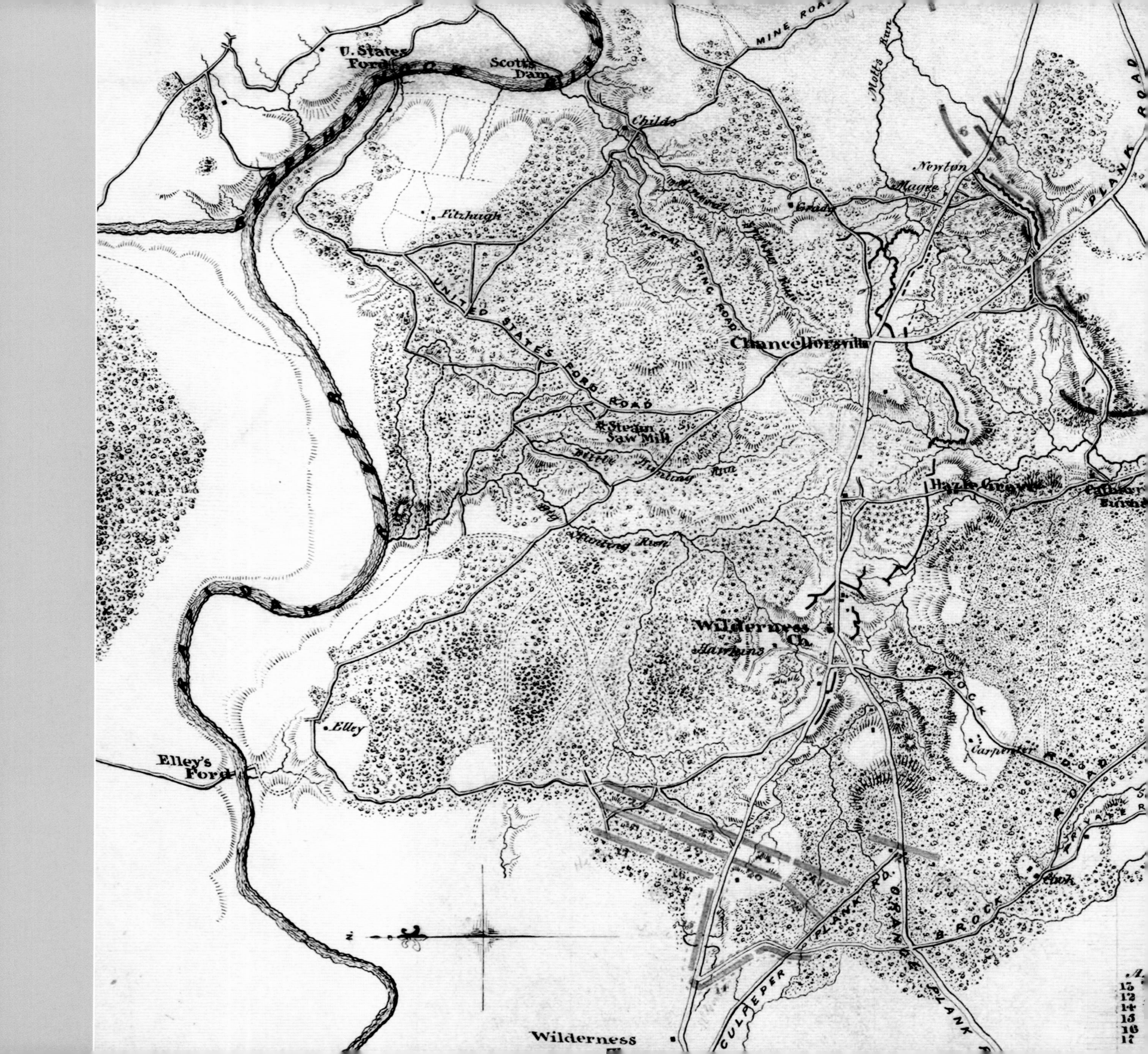

U. States Ford
Scotts Dam
MINE ROAD
Childs
Fitzhugh
Newton
Grady
PLANK ROAD
MINERAL SPRING ROAD
UNITED STATES FORD ROAD
Chancellorsville
Steam Saw Mill
Little Hunting Run
Hunting Run
Hazle Grove
Wilderness Ch.
Hawkins
BROCK ROAD
Carpenter
Elley
Elley's Ford
Cook
CULPEPER PLANK RD.
ORANGE PLANK
Wilderness

The Battle of Chancellorsville, or "THE WILDERNESS" Saturday May 2nd 1863.

0 ¼ ½ 1. 2.

Scale of Miles

Piney Branch or Yellow Ch.

Mine Creek

CATHARPEN ROAD

BROCK ROAD

Todd's Tavern

ision.	TRIMBLE'S Division	D.H.HILL'S Division
ade	18 Nicholl's Brigade	22 Iverson's Brigade
	19 J.M.Jones "	23 Rodes "
	20 Colston "	24 Doles "
	21 Paxton's (2nd Va "	25 Colquitt's "
		26 Ramseur's "

Battle of Chancellorsville [Map #132]

Although Hotchkiss titled his sketch "Wilderness Campaign," he dated the document May 2, 1863, which corresponds with the movements of Jackson's Second Corps during the battle of Chancellorsville. Jackson's route led through concealed, roundabout roads in the baffling Wilderness area. The thickly wooded area of oak trees and scrub brush became known as the Wilderness because one could not stray far from the road without the risk of becoming lost. Jackson formed his brigades (in red) during the late afternoon and launched a surprise assault on Hooker's right flank.

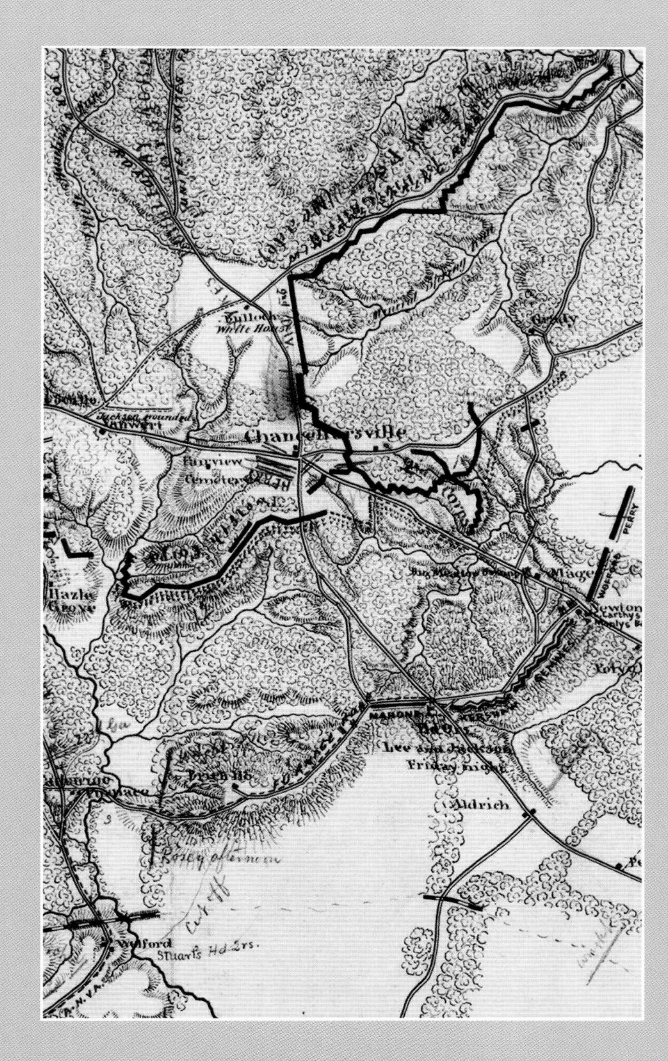

Battle of Chancellorsville [Envelope Map #131]

During the winter and spring of 1863 when Hotchkiss surveyed and sketched much of Spotsylvania and Stafford counties, he took particular interest in the puzzling Wilderness area. Although Stuart's cavalry was familiar with roads leading to the many fords along the Rappahannock and Rapidan rivers, the only current maps were the ones produced by Hotchkiss. His Chancellorsville map shows the concentration of Federal forces around the town and the positions of Confederate forces before General Jackson launched his assault on Hooker's right flank.

A short distance away on the Orange Plank Road, Howard heard the sound of fighting and emerged from his headquarters at Dowdall's Tavern. Deer loped by as he rode to a ridge to investigate. From there he observed the First Infantry Division rushing toward him in panic. He grabbed a Union flag, waved a revolver in his other hand, and shouted, "Halt! Halt! I'm ruined, I'm ruined," but the terrified blueclads sped by him without stopping to catch their breath.

One reserve regiment, the 154th New York, managed to slow the Confederate advance, but Trimble's division, following behind Hill's division, ran through the New Yorkers and swept by Dowdall's Tavern. General Sickles heard the sound of fighting and began to send blueclads from the Third Corps to support Howard. Federal artillery parked in Hazel Grove rotated 180 degrees and began lobbing shells into the Confederate brigades descending on Chancellorsville, slowing Confederate momentum.

Hooker had been enjoying an evening breeze from the porch of the Chancellor House, and due to an aberration of the wind, he did not hear the firing. Suddenly a commotion down the pike reached the ears of Captain Harry Russell. He dashed from the porch, looked down the road at a wave of panic-stricken soldiers, and shouted, "My God! Here they come!" Hooker jumped on his horse but failed to stop the flight of the soldiers, all from his old division from the Third Corps. As darkness closed over the Old Turnpike, fighting slowly subsided.

Lieutenant James Keith Boswell (1839–63) *became Stonewall Jackson's chief engineer in 1862 and soon became Hotchkiss's closest friend and tentmate. The two men traveled together, Boswell performing engineering duties while Hotchkiss sketched maps and shared surveying responsibilities with his friend. Boswell's death at Chancellorsville saddened Hotchkiss interminably.*

JACKSON'S FATAL RECONNAISSANCE

Darkness forced Jackson to curtail the assault about a mile from Chancellorsville. Having thrown Federal forces into utter chaos, he yearned to finish the work. The general had been following his brigades through the woods, pressing them forward and cheering every small victory. He ordered A. P. Hill's division to relieve Rodes and prepare for a night attack. Jackson did not have Hotchkiss nearby to send on a reconnaissance, and all the terrain was unfamiliar, made doubly so by shadows cast by bright moonlight. He advanced slowly on his horse with several members of his staff in an effort to locate enemy lines. When he heard blueclads ahead felling trees for defensive works, he turned back.

As Jackson approached his own lines, skittish pickets from the Eighteenth North Carolina mistook the general's party for Federal cavalry and opened fire. Nobody informed the Carolinians that Jackson would be returning their way from a reconnaissance. Several men in the reconnaissance party fell wounded from their mounts. When an aide shouted, "You are firing on your own men!" a picket shouted back, "It's a lie! Pour it into them, boys!" The second volley raked Jackson's group. Bullets hit the general twice in the left arm and once in the hand. His horse bolted through the woods, and Jackson's head struck a branch before he managed to steady his horse. Finding the general still mounted on a side road, Dr. Hunter McGuire examined Jackson's arm and put him on a litter to be carried to the rear. McGuire, a personal friend and Jackson's staff medical officer, told the general that the left arm would have to be amputated. Jackson consented and turned his corps over to A. P. Hill. When he learned that Hill had also been wounded, he handed the command to Jeb Stuart.

On a late-night reconnaissance during the battle of Chancellorsville, Stonewall Jackson suffered a mortal wound.

Chief Corps Surgeon Hunter Holmes McGuire (1835–1900) *joined Jackson's staff as a private in 1861 and remained on the general's staff until becoming medical director for the Army of Northern Virginia. Hotchkiss maintained a friendship with McGuire that lasted into the 1890s.*

Hotchkiss went immediately to Wilderness Tavern on learning of Jackson's injury. He spoke with Dr. McGuire, who explained the situation but said the general would recover. Though absolutely exhausted, Hotchkiss remounted his fatigued horse and rode four hours back to Lee's field headquarters. The general looked both concerned and relieved after Hotchkiss related his conversation with McGuire.

On May 10 Jackson died from complications and pneumonia brought on by his injuries. That morning, on learning that Jackson was dying, Lee sorrowfully lamented, "He has lost his left arm, but I have lost my right."

HOTCHKISS'S JOURNAL: MAY 3, 1863

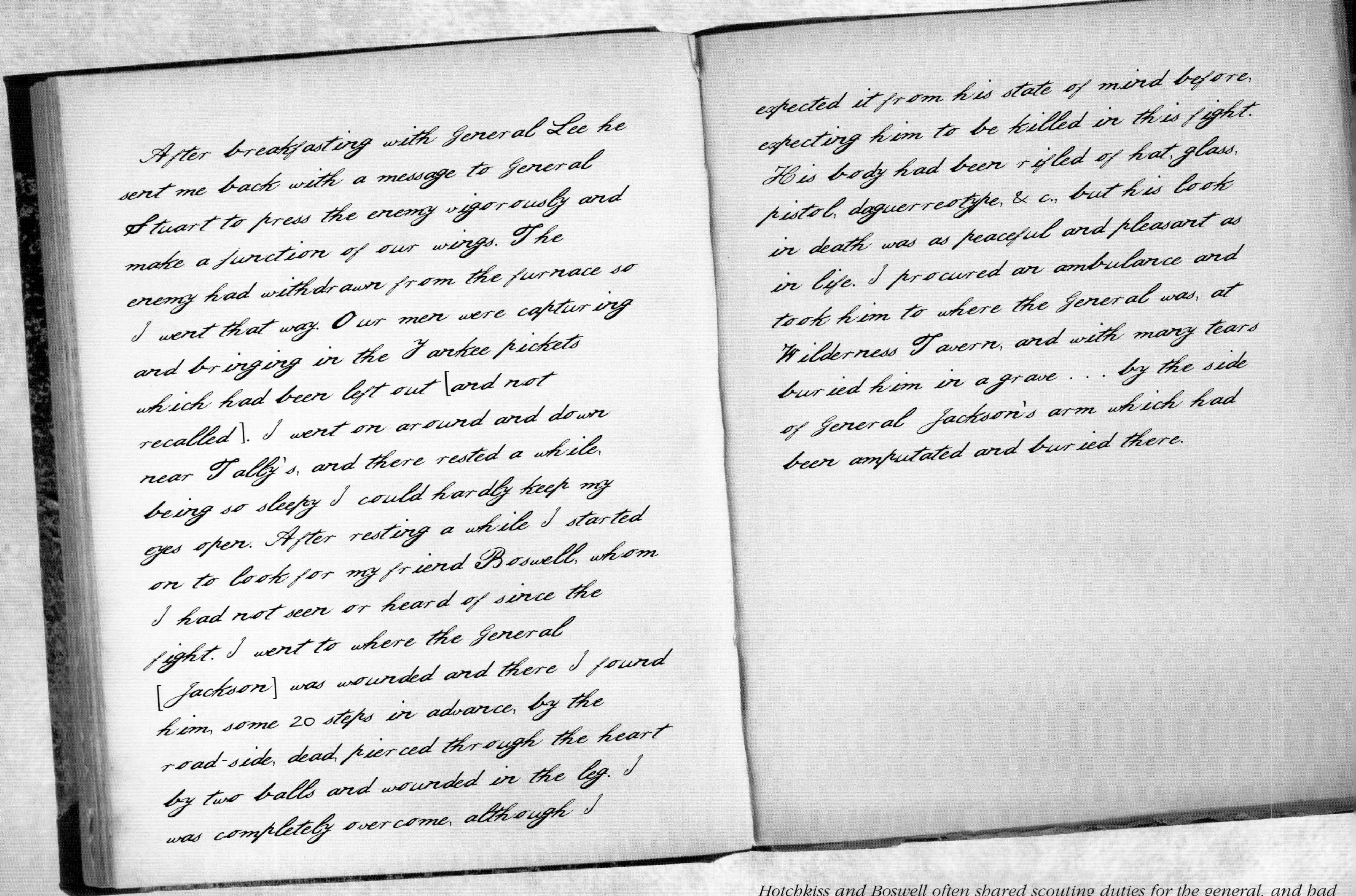

After breakfasting with General Lee he sent me back with a message to General Stuart to press the enemy vigorously and make a junction of our wings. The enemy had withdrawn from the furnace so I went that way. Our men were capturing and bringing in the Yankee pickets which had been left out [and not recalled]. I went on around and down near Tally's, and there rested a while, being so sleepy I could hardly keep my eyes open. After resting a while I started on to look for my friend Boswell, whom I had not seen or heard of since the fight. I went to where the General [Jackson] was wounded and there I found him, some 20 steps in advance, by the road-side, dead, pierced through the heart by two balls and wounded in the leg. I was completely overcome, although I expected it from his state of mind before, expecting him to be killed in this fight. His body had been rifled of hat, glass, pistol, daguerreotype, &c., but his look in death was as peaceful and pleasant as in life. I procured an ambulance and took him to where the General was, at Wilderness Tavern, and with many tears buried him in a grave . . . by the side of General Jackson's arm which had been amputated and buried there.

Hotchkiss and Boswell often shared scouting duties for the general, and had Hotchkiss not returned to Lee's headquarters with messages the previous evening, the body lying dead in the Wilderness might have been his own.

THE CHANCELLORSVILLE CAMPAIGN: MAY 3–4, 1863

At dawn on May 3, General Stuart launched a frontal attack on Hooker's forces, which during the night had been spread about the fields at Chancellorsville. Rodes's battered division had fallen back to Dowdall's Tavern to reorganize, and Brigadier General Henry Heth had assumed command of A. P. Hill's division. The lopsided lineup of forces gave the Federals a huge advantage. Hooker had compressed Sickles's Third Corps, Slocum's Twelfth Corps, and Couch's Second Corps around the town with Meade's Fifth Corps and Howard's Eleventh Corps on two key roads leading to the fords. West of Chancellorsville, Stuart had only Heth's division, Colston's division, and Rodes's exhausted division in reserve to assault the blueclads. Lee wanted to drive Hooker back across the river and risked Anderson's and McLaws's divisions on the left of the Federal line as a show of force. He intended to squeeze Hooker's forces in a vise between Stuart, McLaws, and Anderson. When Dorsey Pender's brigade broke through the Federal line a short distance from Chancellorsville, Hooker became aware of other Confederate units closing on his flanks and began to worry. Although a determined Federal counterattack would probably have broken the Confederate assault, Hooker chose to fight on the defensive.

As the battle ebbed and flowed in the area around Chancellorsville, a wayward Confederate shell fired from Brigadier General James J. Archer's batteries on Hazel Grove struck a pillar on the porch of the Chancellor House where Hooker stood. The pillar split and part of it hit Hooker on the head, knocking him unconscious. Thinking Hooker was dead, officers began moving the general to the backyard for burial. General Couch, next in command, arrived at the Chancellor House without any knowledge of the military situation, and while contemplating what to do with the army, Hooker woke up with a severe concussion and part of his right side paralyzed. Couch promptly departed, and Hooker attempted to get on his horse. Another shell struck the blanket where Hooker had been lying, and perhaps trusting in omens, he worried the next missile would take his life. Despite his condition and the inability to think clearly, Hooker refused to relinquish command.

Around 9:00 a.m., Hooker finally realized he could not lead, summoned Couch from the field, and told him to fall back to the Rapidan. Lee could see the Federal line buckling as units began scrambling for the fords. He told Stuart, McLaws, and Anderson to keep pressing, and 75,000 Federals withdrew from fewer than 25,000 grayclads.

THE FEDERAL FORCE AT FALMOUTH

On May 4, after crossing the Rappahannock on pontoons at Falmouth, Sedgwick's Sixth Corps finally reached Salem Church on the Old Turnpike. McLaws and Anderson did an about-face and checked Sedgwick's advance. By then, Hooker's force at Chancellorsville had withdrawn. While McLaws and Anderson fought Sedgwick to a standstill, General Early's division arrived from Marye's Heights and struck Sedgwick's flank. Lee intended to destroy Sedgwick's corps, but dusk began to settle over the battlefield before he could get the rest of his forces in place. At 6:45 that evening, Sedgwick pulled his corps closer to the Rappahannock and later that night escaped across the river. By May 6 the entire Army of the Potomac had crossed the river and the Chancellorsville campaign ended in a disastrous defeat for the Union.

The Confederate victory began with Jackson's brilliant flank attack on May 3, an operation that may never have happened had a frazzled mapmaker named Jed Hotchkiss not discovered the route that led to victory.

BATTLE OF CHANCELLORSVILLE CASUALTIES

	Engaged	Killed	Wounded	Missing	Total
Confederate	60,892	1,665	9,081	2,018	12,764
Federal	133,868	1,606	9,762	5,919	17,287

CHAPTER FIVE: ON TO GETTYSBURG

Having lost both the general and Boswell, his best friend, Hotchkiss sent a letter home on May 19, two days after attending Jackson's funeral service. He wrote: "We miss him all the time and a void is made here which time can hardly fill. It seems not like our old Hd. Qrs. to any of us, and less to me than to any one else, for my tent mate [Boswell] is gone as well as my General. I do not know whether I will stay here or not."

For several days Hotchkiss served two masters, General Lee and General Stuart. After Hooker retired across the Rapidan, Lee ordered maps of the area in an effort to find ways to keep the Federals running. Hotchkiss spoke of the fords and roads that led to the enemy's new, fortified position, but Lee said Hooker's defenses were too formidable and finally decided to rest the army. He asked for a map of the Chancellorsville battlefield (see envelope), and for the next four weeks Hotchkiss borrowed men to assist with surveying. With Howell Brown, a member of his staff, he began to re-create the landscape of the campaign. He interviewed Confederate officers and captured Union officers to pinpoint the placement and movement of forces during the two-day battle. As Hotchkiss had erected his tent near headquarters, many of Lee's generals interrupted his work because they wanted maps of their own.

As June approached, Hotchkiss observed Lee's impatience. From a new flurry of requests for maps of northern Virginia, Hotchkiss predicted the army would move north into unfamiliar areas.

On May 23, 1863, General Ewell returned to duty after losing a leg at Antietam and assumed command of the Second Corps. Ewell served under Jackson during the Shenandoah Valley campaign and recognized the importance of Hotchkiss's work. Longstreet returned from the Suffolk area and resumed command of the First Corps. Lee's command swelled to 89,000 men, which he considered too many for two corps. He elevated A. P. Hill to lieutenant general and placed him in command of the newly created Third Corps. General Stuart remained in command of the 10,000-man Cavalry Corps, and Brigadier General William N. Pendleton retained the title chief of artillery, though most of his batteries had been distributed among the divisions. On June 1 Ewell officially took command of the Second Corps and retained most of Jackson's staff, including Hotchkiss. The general received hearty cheers, but the entire corps privately questioned whether their new leader would have the drive, stamina, and tactical finesse of "Old Jack."

THE MARCH BEGINS

June 4 was an extremely hectic day for Hotchkiss. Lee wanted maps prepared of routes to Culpepper, and staff officers from other commands asked for maps of routes beyond Culpepper. The following day, Ewell struck camp and invited Hotchkiss to ride with him. As the ambulance rumbled along the Old Turnpike, Hotchkiss gave the general a running account of the battle of Chancellorsville. On June 7 Hotchkiss reached Culpepper and found the entire army in the area. Since taking command of the Second Corps, Ewell received his first rousing cheer, the same kind, wrote Hotchkiss, "as had been their habit with General Jackson." Ewell's Second Corps consisted of three divisions commanded by Robert Rodes, Jubal Early, and Edward Johnson, who had recently returned to duty after suffering a wound on May 8, 1862, during the battle of McDowell.

Word spread that Lee intended to take the army into the valley. Ewell asked for maps showing the routes to Front Royal, as he wanted to march at first light. Hotchkiss had mapped the entire Shenandoah Valley, so he merely added routes from Culpepper to the gaps in the Blue Ridge.

In early June, Hooker suspected Lee of undertaking a movement and began probing Confederate positions. At Franklin's Crossing on June 4, part of Sedgwick's Sixth Corps crossed the Rappahannock near Deep Run but were repulsed by Hill's grayclads firing from rifle pits. Having tested one flank of Lee's army, Hooker waited five days and on June 9 sent a two-pronged reconnaissance in force across Beverly Ford and Kelley's Ford near Brandy Station. General Pleasonton's unexpected attack caught Stuart's cavalry off guard. Ewell sent Rodes's division, which stopped short of Brandy Station on learning that Stuart had regained control of the situation. Pleasonton withdrew without observing Lee's forces at Culpepper, less than eight miles away. Pleasonton's attack gave Stuart a fright and gave Lee concerns. Stuart's role during the upcoming campaign would be to screen Lee's army, which he barely succeeded in doing at Brandy Station. Lee suspected that Pleasonton's scouts had probably seen at least some part of the Confederate force at Culpepper and put the army in motion.

On June 10, "Rodes started for Front Royal in the p.m.," wrote Hotchkiss. "Johnson and Early took the

Sperryville Turnpike." Hotchkiss joined the procession the following day. Lee wanted the Federal garrisons at Winchester and Martinsburg neutralized, and gave the task to Ewell.

THE UNION FORCES

From Chancellorsville to Fredericksburg, Hooker commanded 112,000 effectives, but Lincoln had ordered him to not engage in a major campaign without permission from Washington. General Schenck's Seventh Corps guarded the Shenandoah Valley with Benjamin F. Kelley's 10,000-man division at Harpers Ferry, Benjamin F. Smith's 1,200-man brigade at Martinsburg, Robert H. Milroy's 9,000-man division at Winchester, and Colonel Andrew T. McReynolds's 1,800-man brigade at Berryville.

When General Halleck learned that Ewell's corps had passed into the valley, he ordered Milroy to drop back and join forces with Kelley at Harpers Ferry. Milroy insisted that he could hold Winchester and refused to withdraw. On June 14, Lincoln wired General Schenck, Milroy's superior, and said, "Get Milroy from Winchester to Harpers Ferry . . . If he remains he will get gobbled up, if he is not already past salvation." Confederates were already blocking the roads to Berryville and Martinsburg, and the strangulation of Winchester had begun.

Hotchkiss provided maps that would take Rodes's division through Ashby's Gap and Snicker's Gap while Johnson's and Early's divisions entered the valley through Chester Gap at Front Royal. Screened by Brigadier General Micah Jenkins's Confederate cavalry, Ewell expected Rodes to march through Berryville, bypass Winchester, and go straight down the pike to Martinsburg, threatening Milroy's communications and line of retreat. On June 13 Federal scouts spotted Jenkins's cavalry approaching Berryville with Rodes's infantry and warned

RICHARD STODDERT EWELL (1817–72)

Although born in the District of Columbia in 1817, "Dick" Ewell grew up in Prince William County, Virginia. He graduated from West Point in 1840, thirteenth in a class of forty-two, and joined the First U.S. Dragoons. Brevetted captain during the Mexican War, Ewell returned to the frontier and served in the West until the outbreak of secession. He resigned his commission from the U.S. Army on May 7, 1861, and joined the Virginia militia.

Ewell began his career during the Civil War as a colonel in charge of a cavalry instruction camp. Promoted to brigadier general on June 17, 1861, he commanded an infantry brigade at First Manassas. Elevated to major general on January 23, 1862, Ewell served admirably under Stonewall Jackson in the Shenandoah Valley. Throughout the campaign, he grumbled profusely over Jackson's patent secrecy, which on several occasions prevented him from issuing timely orders to his subordinates. By then, Ewell had lost most of his hair, and his men surreptitiously referred to him as "Old Bald Head." Ewell's long-beaked nose, coarse shaggy whiskers, and a pair of glinting eyes that darted nervously from one object to another reminded his staff of a person out of harmony with his surroundings.

Wherever Jackson went, Ewell followed. He led his division during the Seven Days battles, Cedar Mountain, and Second Manassas, where he lost his leg at Groveton. After Jackson was mortally wounded at Chancellorsville, Ewell returned to the Army of Northern Virginia to take command of Stonewall's venerated Second Corps. Hotchkiss remembered Ewell as being a dependable and relentless fighter while serving under Jackson. When Ewell appeared in camp on June 1, 1863, to take command of the corps, Hotchkiss noted in his journal, "General Ewell looks feebly." Ewell had always been a confirmed bachelor, but he married after losing his leg. He brought his wife to camp and introduced her as "Mrs. Brown" or "the widow Brown," as if she was still an appendage of her former husband.

Having an artificial limb, Ewell could not mount a horse without help. In the upcoming battles of the Army of Northern Virginia, he often rode in an ambulance. Whether or not the loss of a leg took some of the fight out of Ewell remained to be determined as General Lee prepared for the next campaign. Ewell, however, admired and depended on Hotchkiss, and the two men established a friendship that extended for a long time after the war.

McReynolds. Because Jenkins had failed to screen Rodes properly, McReynolds successfully extricated his small force from Berryville and joined Milroy at Winchester.

About 9:00 a.m. on June 13, after the Stonewall Brigade drove off Federal pickets, Ewell turned to Johnson and said, "You are the operator now, I am only a looker-on." Johnson sent skirmishes ahead to test the enemy's strength along Hogg Run, which crossed the Berryville road about two miles south of Winchester. Artillery opened around noon and firing began to spatter along the Valley Turnpike in the vicinity of Kernstown.

From the hill on Millwood Road, Ewell spotted two fortified earthworks and an unfinished works that lay beyond, all three northwest of Winchester. He correctly deduced that Main Fort (also called Flag Fort or Milroy's Fort) dominated Star Fort, located on a ridge north of Main Fort. The occupied but unfinished West Fort faced Little North Mountain. Ewell ordered Early to advance through the woods and assault West Fort while John Gordon's brigade demonstrated from the south and Johnson's division threatened Winchester from the east.

Early concealed his march by working around woods and ridges west of Winchester. The column passed behind Little North Mountain and mounted artillery on the ridges. Although Milroy had spent much of the day hoisted in a basket at Main Fort, he never spotted Early's movements until the late afternoon when Confederate batteries opened on West Fort. Confederate infantry emerged from the surrounding woods and began to encircle Winchester. Caught by surprise and suddenly concerned about a possible envelopment, Milroy responded with Federal artillery. Ewell ordered up more batteries and, with shells descending from all directions, quieted the Federal guns.

At 6:30 p.m., Brigadier General Harry T. Hays's Louisiana Brigade broke a temporary lull in the action with a shrieking rebel yell. More grayclads burst from the woods along Little North Mountain and assaulted West Fort. Ewell moved to a closer ridge to watch the action and tumbled over when struck in the chest by a spent minié ball. Badly bruised but not injured, Ewell resumed his position and from a distance cheered as Hays's infantry captured the fort and, as darkness approached, turned the Federal guns on the fleeing blueclads.

Although neither Main Fort nor Star Fort suffered much damage, Ewell anticipated that Milroy would attempt to escape during the night. He sent three brigades from Johnson's division on roads east of Winchester to block the Valley Pike to Martinsburg and the railroad cut to Stephenson's Depot. Now convinced he was being surrounded, Milroy spiked his guns shortly before midnight and ordered a general evacuation.

Riding with his staff east of Winchester around 3:00 a.m., Johnson heard voices and the sound of nickering horses on the pike and in the railroad cut. Ewell had been right—Milroy was escaping. Johnson had only 3,500 men to oppose up to 9,000 Federals, but he knew that Brigadier General James Walker's brigade was approaching from the rear.

When Johnson opened fire at 4:00 a.m., Milroy squandered his numerical advantage by attacking piecemeal. The blueclads butted up against Brigadier General George H. "Maryland" Steuart's grayclads, who had taken a strong position backed with artillery in the railroad cut. By the first streaks of dawn, Federal dead and wounded began to pile up on the tracks. Milroy sent in another brigade, and Confederate cannon charged with grapeshot and canister mowed down more blueclads. Johnson's men had nearly exhausted their ammunition when Walker's brigade arrived. The Federals panicked, broke formation, and scattered in every direction.

Second Battle of Winchester [Map #155]

"Many people came out to see the fight," wrote Hotchkiss, who during other battles usually remained in the rear. Having both his sketchbook and pencils at hand, he observed the opening battle from "the hill on the Millwood Road." He also acted as one of Ewell's couriers, carrying messages to frontline commanders. "Gen. Ewell called me, at an early hour," Hotchkiss wrote on June 16, "to go and make a map of the battlefield, so I was busy all day riding along the lines of the works of the enemy, our lines of march, &c." Months passed before Hotchkiss finished his map.

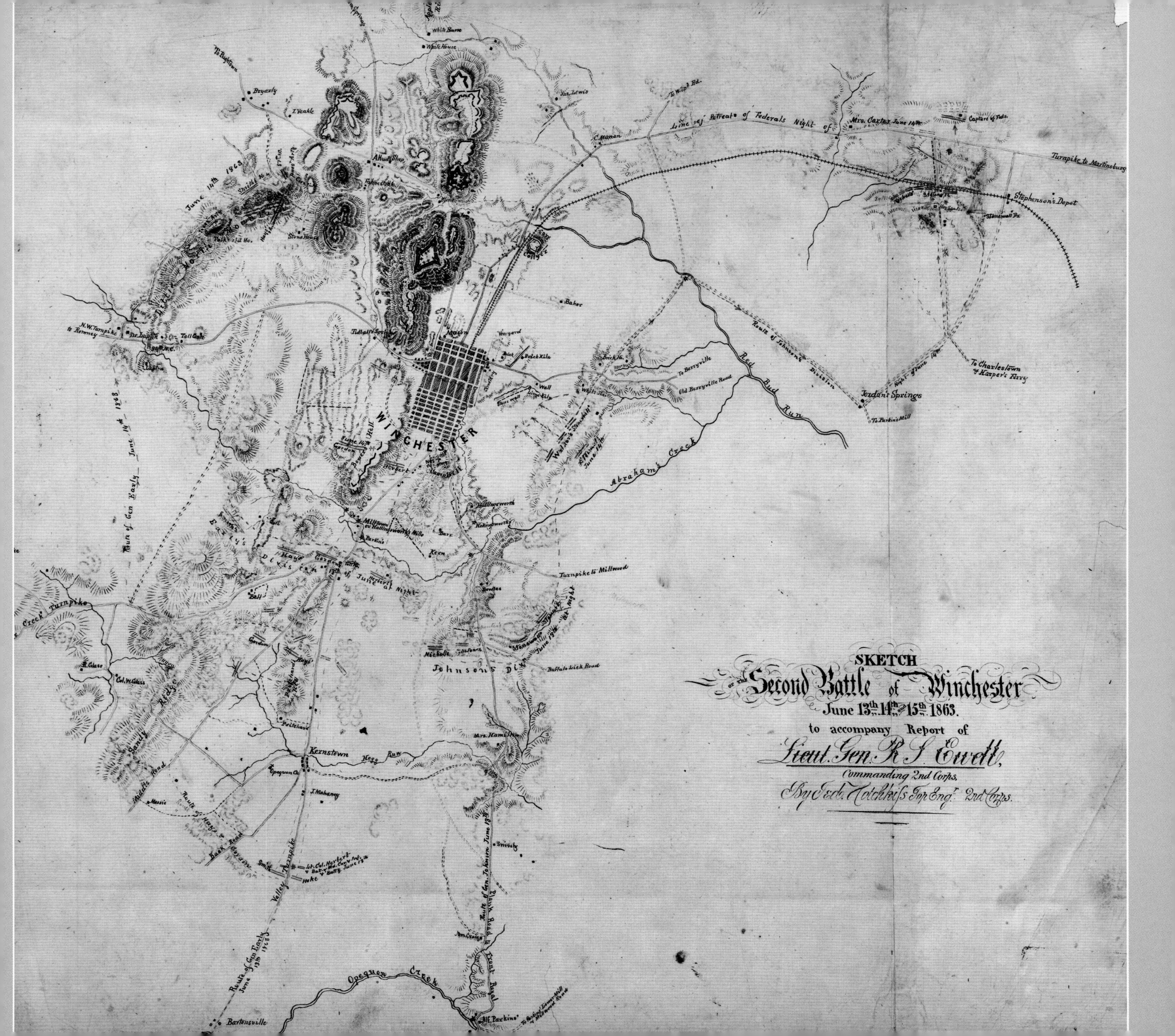
SKETCH
of the Second Battle of Winchester
June 13th, 14th, & 15th, 1863.
to accompany Report of
Lieut. Gen. R. S. Ewell,
Commanding 2nd Corps,
By Jed. Hotchkiss Top. Eng. 2nd Corps.
WINCHESTER
Stephenson's Depot
Turnpike to Martinsburg
Line of Retreat of Federals Night of
Jordan's Springs
To Charlestown & Harper's Ferry
Route of Johnson's Division
Red Bud Run
Abraham's Creek
To Berryville
Old Berryville Road
Turnpike to Millwood
Buffalo Lick Road
Johnson's Div.
Kernstown
Hogg Run
Opequon Ch.
Opequon Creek
Bartonsville
Valley Turnpike
Back Road
Sandy Ridge
N.W. Turnpike to Romney
Run of Gen Early June 14th 1863
June 14th 1863
Baker
Vineyard
Brick Kiln
Millwood
Kern
Bowles
Nicholls
Mrs. Hamilton
Snively
Route of Gen. Johnson June 13th
Front Royal Road
Route of Gen Early June 13th 1863
J. Mahaney
Massie
Col. W. Glass
Toll Gate
Dr. Dunlop
Brearly
T. Yeakle
White House
White Barn
Jas. Lewis
Baker
Tidball's Spring
Parkins
Abraham's Creek
To Romney
Turnpike

AMBROSE POWELL HILL (1825–65)

Born at Culpepper, Virginia, in 1825, A. P. Hill graduated from West Point in 1847, fifteenth in a class of thirty-eight, and joined the artillery. He served briefly in the Mexican War and fought in the Third Seminole War. On March 1, 1861, Hill came off the frontier, resigned from the U.S. Army, and became colonel of the Thirteenth Virginia Infantry.

Appointed brigadier general on February 26, 1862, Hill's ferocious fighting at Williamsburg on May 5 earned him command of a division and a promotion to major general. After the rapid movements of his division during the Seven Days battles, where he spearheaded several attacks and fought fiercely, the unit became known as "Hill's Light Division." At Cedar Mountain, while serving under Jackson, Hill's division earned the reputation of being one of America's great fighting units. Hill's Light Division probably saved Lee's army at Antietam, and at Fredericksburg the division repulsed the Federal attack on the Confederate right. Jackson and Hill did not always agree, but at Chancellorsville Jackson put Hill's division in the forefront of the assault on the Federal right flank. When Jackson suffered a mortal wound, he transferred command of the corps to Hill, who was also wounded.

While serving under Jackson, Hill exhibited a side of his personality that troubled General Lee. For this reason, Lee hesitated to put Hill in charge of the Second Corps. Jackson had once arrested Hill for negligence, and Hill, always combative, demanded a court-martial to vindicate his name. Lee refused to hear the case and ordered both generals to attend to their commands. Hill remained resentful, which caused Lee to question whether his fiery division commander could lead a corps. Lee also worried that Hill, if placed in charge of the Second Corps, would dissolve Jackson's staff and insist upon his own. The Second Corps had one of the most experienced and efficient staffs in the Confederate army, but they sided with Jackson in his feud with Hill. Hotchkiss reflected the sentiment of Jackson's staff, who were all devout like their general, when he wrote, "He [Hill] is not a man of God like Gen. J. and wears not 'the sword of the Lord and of Gideon.'"

Lee avoided the problem by putting Ewell in command of the Second Corps and organizing the Third Corps under Hill. How the experiment would play out in the next campaign would be determined very soon.

Ewell passed his first test as a corps commander with brilliant results. At the cost of only 269 killed and wounded, the Second Corps had inflicted 1,085 casualties, taken 3,358 prisoners, and collected 300 wagons and twenty-three guns. Hotchkiss went directly into Winchester and entered Milroy's headquarters at the Logan house, looking "for paper and engineering materials," which in the Confederacy had become scarce articles. Some might call the activity looting, but Hotchkiss called it necessary.

THE MARCH RESUMES

On June 15, as Ewell rounded up Federal prisoners at Winchester, the Army of Northern Virginia stretched from Rodes's brigade at Williamsport, Maryland, to Hill's Third Corps at Fredericksburg. Lee ordered Longstreet and Hill to hurry up. Four days later, Longstreet passed through the Blue Ridge at Ashby's Gap and Snicker's Gap, but he deposited two brigades of 3,000 men at the passes to support Stuart's cavalry and keep communications open. Charged with screening Lee's movements, Stuart encountered unexpectedly heavy pressure from Pleasonton's Union cavalry and fought several sharp engagements in defense of the passes. Because Federal cavalry failed to penetrate the Blue Ridge, General Hooker lost contact with Lee's army.

On June 17 Ewell sent Hotchkiss and Sandie Pendleton to Williamsport because Rodes had moved from Martinsburg into Maryland and needed maps. Most of Hotchkiss's work for Maryland and Pennsylvania came from maps made by others before the war. Until he could update the maps by making his own, he used what he had. Hotchkiss and his staff worked late into the night copying maps to guide the officers who might get lost without them.

The following day, Ewell arrived with Major Gilbert Moxley Sorrel from Longstreet's staff to discuss the movements of the two corps. Aside from Ewell, Sorrel, and Rodes, only Hotchkiss knew what roads the army would likely take. Ewell went with Sorrel to Leetown to confer with Longstreet, and Hotchkiss went to Shepherdstown to discuss routes with Johnson.

On June 26 Rodes crossed into Pennsylvania and marched to Chambersburg, Early moved toward Gettysburg, and the rest of the corps followed. "The people looked sullen," Hotchkiss reported that day, and after he wandered into a roadside orchard, he added more cheerfully, "The cherries are very fine." The lush farms of Pennsylvania reminded Hotchkiss of how the crops and livestock once looked before war ravaged the farms of the valley.

Ewell wanted maps made for movements planned over the next four days, so Hotchkiss went into Chambersburg and purchased paper, county maps, and other supplies. While he laid out routes for Ewell, a courier arrived from Lee with a message predicting that the next major battle would be fought at Gettysburg or Frederick.

With a four-day plan, Ewell put his divisions in motion on June 27 and began marching through southern Pennsylvania. Rodes and Johnson took routes to Carlisle, and Early headed for York. On the following day, Ewell raised the Confederate flag over the U.S. Army barracks at Carlisle while Jenkins's cavalry trotted east to Camp Hill, a small town on the Susquehanna River across from Harrisburg. Hotchkiss intended to follow Rodes to Harrisburg on June 29, but he received a summons from Ewell for maps of Adams County. That afternoon, Ewell received urgent orders from Lee to move toward Gettysburg. On the morning of June 29, Ewell's divisions began to converge on Heidlersburg, twelve miles north of Gettysburg. The four-day plan had expired.

JAMES LONGSTREET (1821–1904)

Born in South Carolina in 1821, Longstreet spent most of his youth on a farm in Augusta, Georgia. When his father died in 1833, he moved with his mother to Somerville, Alabama. Appointed to West Point, Longstreet graduated fifty-fourth in a class of sixty-two that included such notables as Ulysses S. Grant, Henry Halleck, and William T. Sherman, all of whom served with the Union.

Longstreet resigned from the U.S. Army on June 1, 1861, and sought a post in the Confederate army as paymaster. Instead, the war department made him a brigadier general and sent him to Manassas to command the Fourth Brigade. Longstreet deployed and fought his brigade so skillfully at First Manassas that the war department boosted him to major general on October 7, 1861, and placed him in charge of a division.

Longstreet's performance during the Peninsula campaign fluctuated between brilliance and disappointment. His mistakes while leading the right wing at Fair Oaks and Seven Pines cost the Confederacy an early victory. His superb performance during the Seven Days campaign led Lee to put Longstreet in charge of half the army. At Second Manassas, Longstreet's delay in launching a flank attack in support of Jackson's strategic envelopment deprived Lee of the opportunity to crush General Pope's army. Three weeks later, he fought commendably on the defensive at Antietam, received a promotion to lieutenant general on October 9, 1862, and was given command of the newly designated First Corps of the Army of Northern Virginia. Given an opportunity for advancement to full general in the weeks prior to the Chancellorsville campaign, Longstreet took two divisions into southern Virginia and demonstrated an inability to function independently.

Lee observed Longstreet's genius for fighting on the defensive and his sporadic lack of aggressiveness when fighting on the offensive, especially if great risks were involved. When Longstreet rejoined the Army of Northern Virginia after Chancellorsville, he argued against Lee's plan to invade the North. Although Lee had great confidence in Longstreet's fighting ability, referring to him as his "Old War Horse," a great gap existed between Lee's confidence in the strategic offensive as opposed to Longstreet's dependence on a tactical defense.

When the Army of Northern Virginia marched into the Shenandoah Valley on June 5, 1863, to begin Lee's summer offensive, Longstreet erroneously assumed that his commander intended to fight on the tactical defensive.

The March Resumes [Map #158]

At a later date, Hotchkiss sketched the routes followed by General Ewell's Second Corps from Grace Church, near Fredericksburg, Virginia, to Harrisburg and Wrightsville, Pennsylvania. The red line plots the route north while the green line plots the routes followed by Ewell when the Army of Northern Virginia withdrew from Gettysburg on the night of July 4 and returned to Orange Court House on August 1, 1863.

THE UNION DILEMMA

During the month of June, Hooker lived through each day in a conundrum, completely puzzled by Lee's movements. On June 13 he sent part of his army north, presumably to follow a route parallel to Lee but on the eastern side of the Blue Ridge Mountains. He knew that some part of Lee's army had moved down the Shenandoah Valley because of Milroy's defeat at Winchester. Not until June 25, when Ewell's corps moved into Pennsylvania and shocked the North, did Hooker react. He crossed the Potomac at Leesburg and moved into Maryland.

Lincoln, Stanton, and Halleck had already decided to relieve Hooker after Chancellorsville but waited for a situation that would induce the temperamental general to resign. The long-awaited event occurred on June 27 when Hooker demanded troops from the garrison at Harpers Ferry and Halleck refused. Hooker accommodated the administration by indignantly resigning, and on June 28 a staff officer from the war department arrived in Frederick with orders for General Meade to take immediate command of the Army of the Potomac. Meade grumbled and groused, attempted to decline the promotion, but was informed that he had no choice in the matter. "Well," he said tartly, "I've been tried and condemned without a hearing, and I suppose I shall have to go to the execution."

Meade did not know where his predecessor had placed the army and, after speaking with Hooker that morning, concluded that the deposed general had no plan or any clear recollection of the army's dispositions. By the end of the day, Meade had located his forces and designed a strategy. He put the army on the road and on June 29 formed a twenty-mile defensive position near the Pennsylvania border. Meade planned to interpose the Union army between Lee's forces and Washington by moving through Frederick to Harrisburg. This plan changed the following day when Confederates began looking for shoes.

On June 30, after learning that Longstreet's corps and Hill's corps were in the Chambersburg area, Meade sent Brigadier General John Buford's cavalry on a reconnaissance to Gettysburg. As the Federals rode into the Pennsylvania town from the Emmitsburg Road, Brigadier General James J. Pettigrew's brigade of Henry Heth's division from Hill's corps approached Gettysburg in search of shoes. Pettigrew spotted Federal cavalry roaming through town and turned back toward Cashtown. His instructions had been to find shoes, not attract attention. Scanning the fields to the west from the cupola of the local Lutheran seminary, Buford saw the grayclads retiring and sent the information by courier to Meade's headquarters in Maryland. Meade ordered Major General John Reynolds's First Corps and Major General Oliver O. Howard's Eleventh Corps to march to Gettysburg in the morning.

Most of the Confederate officers had slaves to serve them in camp. Hotchkiss also had one. In the days before the battle of Gettysburg, many slaves were not sure which way the army was going.

Before Gettysburg: June 1863 [Map #159]

Because Ewell and his officers knew nothing about the Gettysburg area and were all operating north of the town, Hotchkiss oriented this map with "Gettysburg" spelled upside down. His interest was to show the roads coming into Gettysburg from the north, and he did so by pasting a piece to the top of the original. The red angle east of Gettysburg shows the position of General Johnson's division, which Hotchkiss added later.

THE BATTLE OF GETTYSBURG: JULY 1–3, 1863

The battle of Gettysburg began on July 1 when A. P. Hill sent Henry Heth's division, followed by Dorsey Pender's division, back to Gettysburg to chase off the Federal cavalry and seize a reportedly large store of shoes. Buford recognized the importance of the town because Gettysburg served as a connecting hub for ten roads and a railroad spur. Although greatly outnumbered, Buford dismounted most of his two cavalry brigades and used them as infantry to prevent Heth from securing the town. Though few in number, the Federals were armed with Spencer breech-loading .52-caliber seven-shot carbines, and when heavy fighting began at 10:00 a.m., the blueclads slowed down the Confederate advance.

As Pender's division arrived in support of Heth and began pressing back the blueclads, the first Federal reinforcements from Reynolds's corps began to arrive in support of Buford. What started as a skirmish began to develop at 10:30 a.m. into a severe engagement. Reynolds agreed with Buford that a major battle would be fought at Gettysburg and that Cemetery Hill, directly south of the town, provided the strongest defensive position and must be preserved at all costs for the Union army.

The Confederates captured McPherson Ridge and began pressing toward Seminary Ridge. A stray Confederate bullet killed General Reynolds as infantry brigades on both sides began cutting each other to pieces. The fighting subsided at noon while both armies reorganized. An hour later, three divisions from Howard's Eleventh Corps arrived. Howard formed two divisions on Buford's right and placed the other in reserve on Cemetery Hill. Fighting again intensified as Confederates attempted to consolidate a position on Seminary Ridge.

At 10:30 a.m., Hotchkiss learned an engagement had erupted at Gettysburg. Ewell's corps approached the town from the Harrisburg Road and threatened the exposed Federal right flank. Howard countered with Carl Schurz's and Francis C. Barlow's divisions, which Ewell repulsed and drove back through the town with heavy losses. Although Meade quickly put Sickles's Third Corps and Slocum's Twelfth Corps on roads to Gettysburg, they did not arrive in time to quell Hill's assault, which collapsed the First Corps on Seminary Ridge and drove the blueclad survivors across the pastures and grain fields to Cemetery Hill.

According to Hotchkiss, Lee came on the field about 4:30 p.m. and ordered Ewell to drive the Federals off Cemetery Hill, "if possible." Early's division held the best position for executing the assault, but Ewell had just been thrown from his horse and appeared to be slightly dazed. After recovering his senses, Ewell made every effort to re-form. At 5:00 p.m., he began to move into Gettysburg with Rodes's division about the same time Slocum's divisions began to file onto Cemetery Hill.

General Hancock arrived in the late afternoon with part of the Second Corps and with instructions from Meade to determine whether Gettysburg was a good place to bring on a general engagement. Hancock examined the topography with Howard and Meade's chief engineer, Gouverneur Warren, and reported that Cemetery Hill was an excellent position of natural strength from which to fight. Ewell's decision to not assault Cemetery Hill, and Lee's "if possible" order, remains one of the major controversies of the Civil War. Sandie Pendleton, a former member of Jackson's staff, lamented that Stonewall Jackson would have attacked without hesitation. Ewell would waste men and resources over the next two days in assaults on Culp's Hill and Cemetery Hill, without success.

Although Hotchkiss had purchased maps of Adams County and worked on expanding them, nobody in the Army of Northern Virginia had reconnoitered Gettysburg's terrain, which became the key feature in the battle that followed.

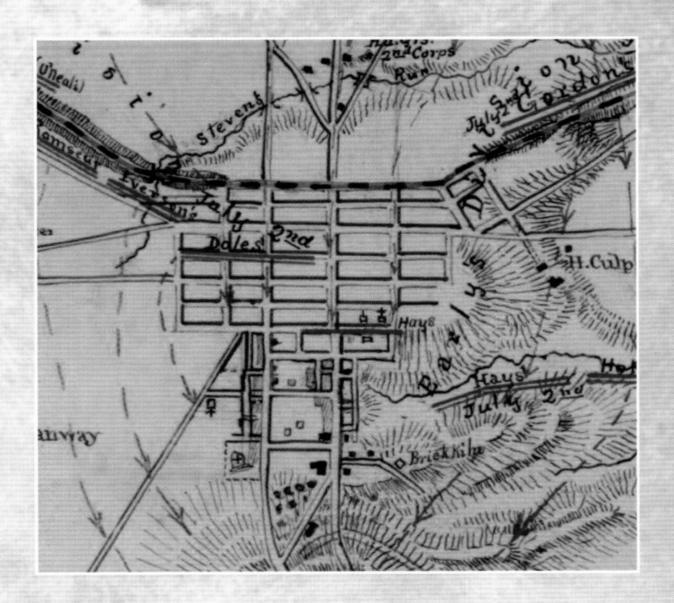

Battle of Gettysburg [Envelope Map #162]

Hotchkiss built this map from a series of sketches, including the one above, to show Confederate positions on July 1 and their subsequent movements into positions taken later in the day. Throughout the day, both armies were constantly in motion. This sketch became the template for the later development by Hotchkiss and Howell Brown of two remarkably accurate and detailed maps produced for the official reports of General Ewell and General Lee.

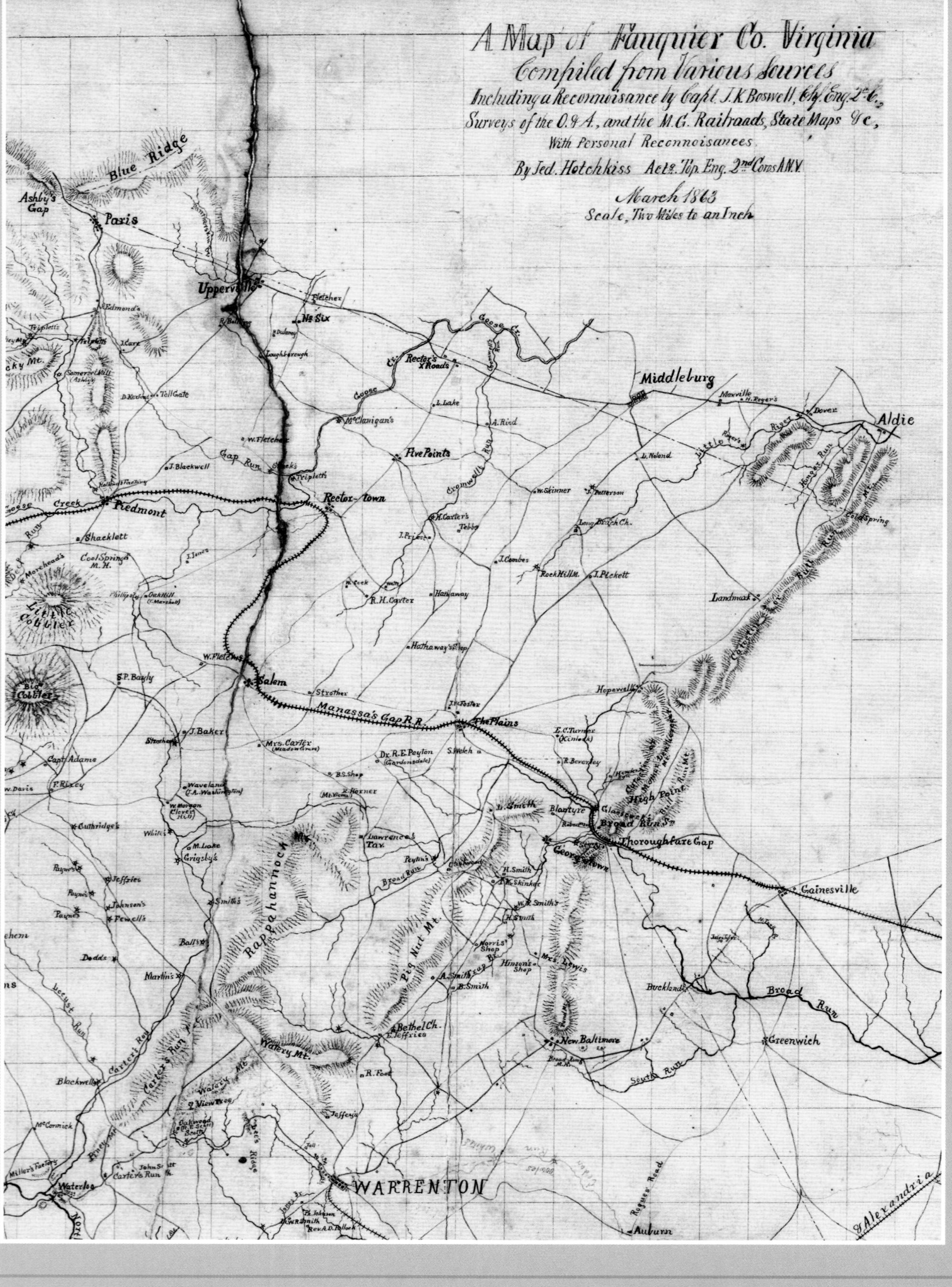

Fauquier County
[Map #31]
Since the battle of Fredericksburg, Hotchkiss had been working on a map of Fauquier County, mainly to have a document that could be used for military purposes. Hotchkiss completed his map in March 1863. Three months later, Jeb Stuart relied on this map during operations in the northern sections of the county.

THE DISAPPEARANCE OF JEB STUART

Until the battle at Brandy Station on June 9, 1863, Stuart's cavalry had become accustomed to having their way with the Northern cavalry. Stuart narrowly escaped defeat at Brandy Station, a battle most Civil War historians agree made the Federal cavalry an equally combative force. Embarrassed by the near disaster, Stuart now felt he had something to prove. A few days later, Lee sent Stuart north to screen the movements of the army in the Shenandoah Valley by preventing Pleasonton's cavalry from penetrating the gaps in the Blue Ridge.

Although minor cavalry clashes occurred throughout Fauquier County, Federal scouting intensified on June 17 when Stuart sent Brigadier General Fitzhugh Lee's brigade to occupy a gap in the Bull Run Mountains at Aldie. Pleasonton approached the gap at 4:30 p.m. with Brigadier General Hugh J. Kilpatrick's cavalry brigade. Mounted and dismounted spirited attacks followed and lasted until dark, after which the Confederates withdrew.

Two days later, Pleasonton's Second Cavalry Division under the command of Brigadier General David M. Gregg struck Stuart at Middleburg. Stuart had destroyed a cavalry regiment from Rhode Island on the previous day, but this time he withdrew about half a mile and took defensive measures.

Pleasonton continued to pursue Stuart, and on June 21 the cavalry clashed near Upperville when Federals attempted to break through Ashby's Gap in the Blue Ridge. Both Stuart and Pleasonton consolidated their widely spread forces for a major cavalry battle. As Federal pressure built on all sides, Stuart sent Brigadier General Wade Hampton's division against Gregg's division, and after several saber charges, Hampton's troopers relieved the pressure. Stuart fell back and successfully defended Ashby's Gap. The three battles at Aldie, Middleburg, and Upperville put Federal losses at 613 and Confederate losses at 510.

Although Stuart accomplished his main mission of screening Lee's march, he took advantage of the general's discretionary orders and on June 24 tried to acquire new laurels by leading three divisions on a wide swing around Hooker's right flank. He cut across the Army of the Potomac's supply route and at Rockville, Maryland, captured 400 prisoners and 125 new army wagons filled with stores. Instead of burning the wagons and paroling the prisoners, he encumbered his march, thereby delaying his return to Lee's scattered forces in Pennsylvania. The long wagon train became the target of Federal cavalry. Stuart fought four cavalry battles that added more delays to his planned juncture with Ewell.

Some of Lee's problematic blunders at Gettysburg, especially during the second day, can be attributed to the absence of Stuart. On July 1, when Lee established his tactical plan for the second day, he did not know the strength of the enemy on Cemetery Hill, nor did he know Meade's dispositions. During the second day's fighting, when it became apparent to Lee that his army was being roughly handled, Stuart arrived. Major Henry B. McClellan, Stuart's adjutant, reported the crisp conversation that followed.

Lee demanded to know where Stuart had been, and said, "I have not heard a word from you for days, and you are the eyes and ears of my army."

"I have brought you 125 wagons and their teams, General," Stuart replied.

"Yes," said Lee. "And they are an impediment to me now."

According to McClellan, Lee recovered quickly from his irritation. "Let me ask your help now," he said. "We will not discuss this matter further. Help me fight these people." But it was already too late.

Major General James Ewell Brown "Jeb" Stuart (1833–64) *led the Army of Northern Virginia's Cavalry Corps until he was killed during General Grant's 1864 spring campaign. Hotchkiss prepared many maps for Stuart and also led the general's scouting parties on numerous occasions.*

GETTYSBURG: THE SECOND DAY

On July 2 Hotchkiss occupied a knoll beside the Hagerstown (Fairfield) Road to watch the battle unfold, and when nothing happened, he began making the rounds with his survey instruments and binoculars to sketch the terrain and troop positioning. Meade had concentrated the Union army on the commanding heights of Cemetery Hill, which when laid out resembled a huge fishhook with the bent tip curving around Culp's Hill, the 2.5-mile shank stretching on Cemetery Hill, and the eye resting on the Round Tops at the southern end. Across fairly open ground to the west, Lee's army spanned out across Seminary Ridge and extended through and around Gettysburg to Johnson's division on Lee's extreme left flank, about seven miles in all. While the Union position was tightly concentrated with up to 122,000 men, the Confederate position of 89,000 men covered twice the area and was partially concentrated. Because Hancock cautioned that the Federal right flank could be turned, Meade put two corps in the vicinity of Culp's Hill and paid less attention to the left flank. In addition to numerical superiority, Meade's army also had the advantage of better observation and open fields of fire.

Hancock was correct in his assumption that the Confederates wanted to attack Culp's Hill, but Ewell informed Lee that it could not be done and recommended that Longstreet attack the Round Tops on Meade's left. Longstreet opposed attacking at all and tried to convince Lee to fight defensively. Lee rejected Longstreet's "tactical defense" theory and ordered him to assault Meade's left flank.

Lee suffered a serious disadvantage by not having Stuart's cavalry available. After screening the passes during Lee's movement north, Stuart went on a reckless campaign that delayed his arrival at Gettysburg and denied Lee the intelligence he needed to fight effectively. Lee therefore lacked information on the terrain around Cemetery Hill and the enemy's dispositions, which impaired his ability to perfect a tactical plan. Meade's left flank was much stronger than Lee realized when he ordered Longstreet to commit two divisions to the assault, placing Hood's force on the far right of the Confederate line and McLaws's division on the near right.

Longstreet's task required that he move Hood's division unseen around the Federal left before assaulting the Round Tops. To take the attention away from Longstreet's assault and prevent Meade from shifting troops to the left, Lee instructed Hill to attack the Federal center in an echelon and Ewell to organize an attack against the Federal right. After failing to attack Culp's Hill on July 1, Ewell agreed to assault the position because he erroneously believed the hill had not been occupied by Federal troops.

What began for Lee as a spirited tactical plan quickly deteriorated into a disjointed operation. The first problem occurred when General Sickles, either ignoring or misunderstanding orders from Meade, sent forward his Third Corps, creating a Federal salient on the battlefield in an area that became known as the Peach Orchard, the Wheat Field, and the Devil's Den. The movement compelled McLaws to forego Longstreet's flank attack and fight Sickles. Hood's division continued to march around the Federal left flank in obedience to Lee's orders but in disobedience to Longstreet's orders. At 4:00 p.m., George T. Anderson's brigade assaulted the Devil's Den, a rocky outcropping at the base of the Round Tops occupied by Brigadier General Hobart Ward's Federal brigade from Sickles's corps. A furious and bloody fight ensued, during which Hood attempted to move Brigadier General Evander M. Law's brigade around the engagement at Devil's Den, drive a small force of Federal skirmishers off Little Round Top, and open a way for Confederates to roll up the Federal left flank.

The tactic almost worked. General Warren observed Confederates climbing Big Round Top and working across the saddle toward Little Round Top. Warren spied Colonel Strong Vincent formed in General Sykes's Fifth Corps' reserve and quickly explained the situation. Without waiting for orders from Sykes, Vincent gathered together his four regiments, double-quicked to Little Round Top, and after a furious battle during which he suffered a mortal wound, contained the Confederate assault. During the fighting, General Hood suffered a wound and permanently lost the use of his arm.

Lee lost the initiative when the flank attack failed, and on that day, nothing went well for the Confederates. What Lee planned as a simultaneous assault on three fronts unraveled when Longstreet, who had resisted Lee's efforts to go on the offensive, delayed the assault in hopes that Lee would then reconsider.

McLaws became engaged in the Wheat Field, Peach Orchard, and Apple Orchard when Sickles unexpectedly pushed forward a salient. Sickles's movement prevented McLaws from joining Hood's flank attack. A. P. Hill failed to vigorously attack the Union center when McLaws became engaged, enabling Meade to shift men and reinforce the Federal left flank.

On the Federal right, Ewell spent most of the day engaged in an ineffectual artillery duel. At 6:00 p.m., a good two hours after Hood's assault on the enemy's left flank, Ewell ordered three divisions to assault Cemetery Hill and Culp's Hill on the enemy's right flank. Ewell's three divisions were separated by

Battle of Gettysburg: The Second Day [Map #43-1]

Lieutenant S. Howell Brown served on Hotchkiss's staff and contributed two maps to the Official Records. *Under Hotchkiss's direction, Brown produced this map for General Lee to show the positions of Federal and Confederate forces on July 2, 1863. As a template, he used Hotchkiss's map from the first day's battle (see envelope), which had also been refined for General Ewell's official report.*

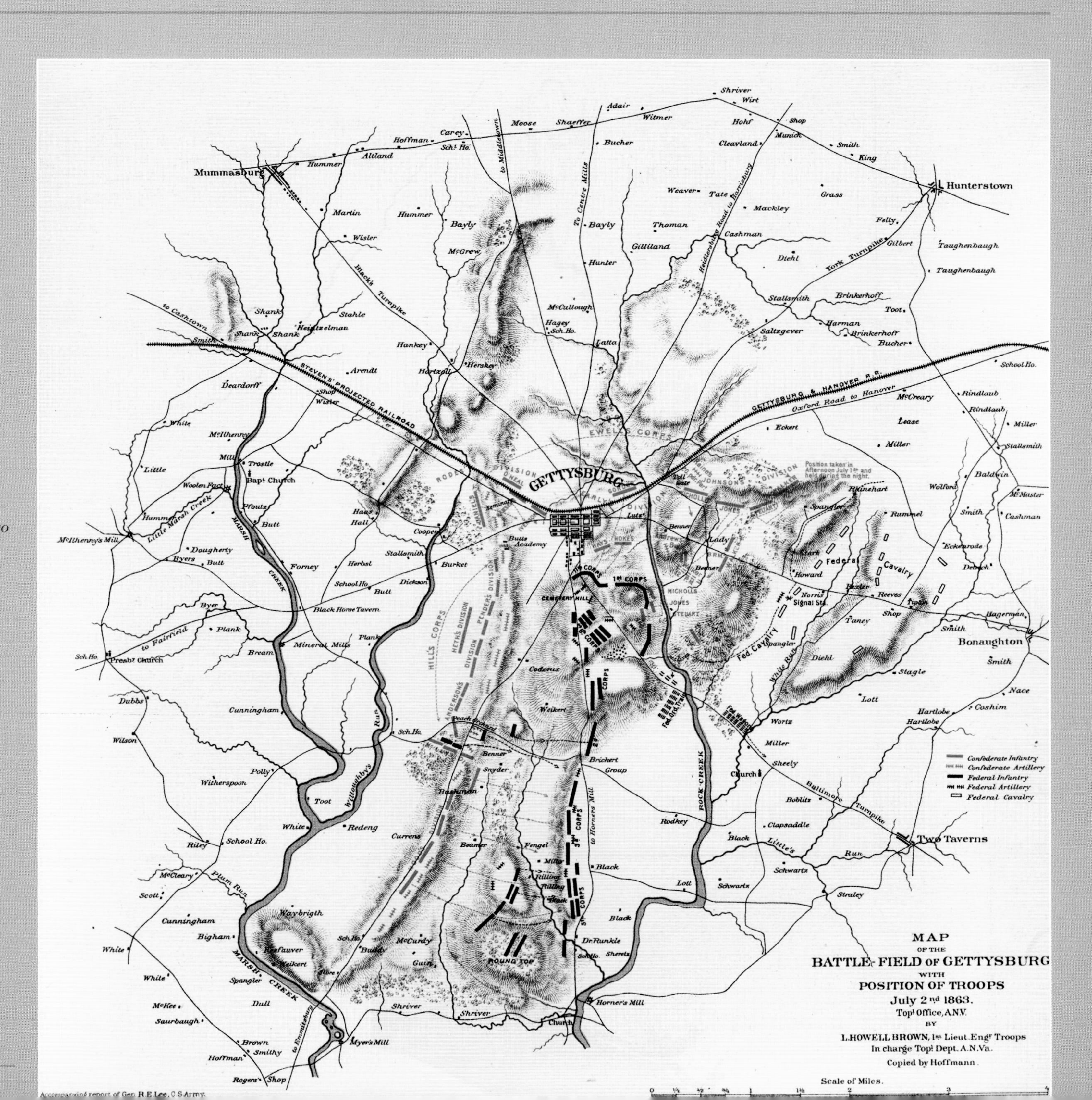

the town of Gettysburg, which resulted in another disconnected attack. Early's division almost broke the Federal line, but withdrew when enemy reinforcements arrived. Darkness interrupted Johnson's attack on Culp's Hill, and when the fighting subsided at nightfall, Lee had thrown most of his divisions against the Union line and gained nothing but casualties. Hotchkiss recalled a conversation early that morning at Ewell's headquarters when Lee arrived to discuss the assault on the Federal right. "He feared we would only take it [Cemetery Hill] at a great sacrifice of life."

Brigadier General Edward Porter Alexander (1835–1910) *commanded General Longstreet's artillery. On the third day at Gettysburg, he bombarded Cemetery Hill for two hours before, in compliance with Longstreet's instructions, he ordered Pickett's Charge.*

GETTYSBURG: THE THIRD DAY

After objecting to Lee's strategy on the second day, Longstreet objected to Lee's plans for the third day as well. Longstreet had been right by arguing for a tactical defense, but Lee had unlimited confidence in the men of the Army of Northern Virginia. Morale remained high despite the previous day's setback, and George Pickett's fresh division had arrived that day from the valley. Lee believed the determined Confederate attacks on July 2 may have unnerved Meade, and they did. During a council of war held that night, Meade considered withdrawing on the advice of most of his generals but ultimately decided to stay and fight.

Lee decided to attack the center of the Federal line with ten brigades after softening the position with 170 guns. He also encouraged Ewell to once more attack Culp's Hill, which would create a diversion. As another distraction, he ordered Stuart to swing the cavalry around Cemetery Hill and assault the Federal rear. Because Lee had unsuccessfully attacked the Federal flanks on the previous day, Meade anticipated the assault on July 3 would focus on the Federal center and strengthen the Confederates' position.

"A terrific cannonade took place about sunrise but with no results," wrote Hotchkiss. "We had 120 pieces of artillery on one ridge. Skirmishing and occasional firing was kept up until 1 p.m." Lee, Hotchkiss, and everyone else on Seminary Ridge were unable to recognize that almost all the shells fired from Confederate batteries soared over the Federal front line and exploded among the reserves and wagons in the rear. Most of the fighting mentioned by Hotchkiss occurred on Culp's Hill, where Johnson's division made no progress after seven hours of brisk fighting that sharply increased the division's casualties.

History inaccurately refers to the next and last big battle at Gettysburg as "Pickett's Charge," probably because Pickett's fresh division was ordered to spearhead the attack. Longstreet commanded the charge, not Pickett. Because Longstreet disagreed with Lee over launching an assault into the teeth of Federal artillery, he deferred the responsibility of forming the brigades for the assault to Pickett. Pickett placed the brigades in such a way that many of his own troops followed rather than led the attack.

Colonel E. Porter Alexander directed the bombardment of Cemetery Hill and, minutes before the barrage was scheduled to commence, he received a puzzling order from Longstreet. Quite unexpectedly, Longstreet abrogated his responsibility for directing the assault. He placed the burden for determining the effect of the bombardment on Alexander and for deciding whether or not the infantry charge should commence. The instruction flew in the face of Lee's orders and flabbergasted Alexander. He protested, but Longstreet insisted. "When our fire is at its best," Alexander replied, "I will advise General Pickett to advance."

At 1:00 p.m., roughly the same time that Johnson's division abandoned the assault on Culp's Hill,

LEE'S TEN BRIGADES

Pettigrew's Division

Marshall Davis Mayo Fry

Pickett's Division

Armistead Garnett Kemper Wilcox

Trimble's Division

Lowrance Lane

Alexander opened with 170 guns. Each gun had between 130 and 150 rounds of ammunition available for the most gigantic barrage in the history of North America. Two hours later, after thinking the guns on Cemetery Hill had been silenced, Alexander scribbled a note to Pickett: "The 18 [Federal] guns have been driven off. For God's sake, come quick, or we cannot support you. Ammunition nearly out." Pickett received the note about 3:00 p.m. He turned to Longstreet and said, "General, shall I advance?" Longstreet glumly lowered his head. Interpreting the nod as affirmation, Pickett replied, "I shall lead my division forward, sir."

PICKETT'S CHARGE

As Alexander's guns fell silent, 13,000 Confederate infantry emerged from the woods lining the base of Seminary Ridge. They marched in formation across half a mile of open field and began to ascend the slope of Cemetery Hill. Federals lined the heights and waited behind breastworks and stone walls for the Confederates to come into easy killing range. While the blueclads waited, Federal artillery opened with shells, blasting gaps in the long gray formations. A few hundred yards from the crest, the Confederate line stopped and formed for the assault. Colonel Birkett D. Fry of Pettigrew's division pointed to a small clump of trees on the Federal line known as "the Angle." As Fry led his men toward those trees, the other nine Confederate brigades guided on him.

As the grayclads advanced, Federal batteries changed from shells to canister, knocking down three or four Confederate soldiers with every round. Pickett's men closed the gaps, maintained a steady formation, and relentlessly approached up and across the gently rising slope. Except for two brigades on the left that began to falter, the others edged toward the center and the clump of trees at the Angle. As Confederates crossed the Emmitsburg Road, Federal infantry opened with a lethal hail of musket fire. Five Confederate brigades, with two following in reserve, charged into point-blank fire as thousands of blueclad reinforcements rushed to the crest and poured enfilading fire into Pickett's rapidly disintegrating line. A few determined Confederates miraculously pierced the Federal position and gained a moment of glory. The effort of those valiant few rightly marked the "high tide of the Confederacy." Pickett's charge stopped in its tracks and began to recede, leaving the battlefield covered with Confederate dead and wounded.

Although Pickett returned from the battle unscathed, three of his brigade commanders fell during the charge. General Garnett, who once

Major General George Edward Pickett (1825–75) *became immortalized at Gettysburg for his famous charge, the failure for which he was blamed. Although the assault was under the direction of General Longstreet, who disagreed with the attack, Longstreet always deflected the blame to Pickett.*

BATTLE OF GETTYSBURG CASUALTIES

	Engaged	Killed	Wounded	Missing	Total
Confederate	75,000	3,903	18,735	5,425	28,063
Federal	88,289	3,155	14,529	5,365	23,049

commanded the Stonewall Brigade, was killed on the field and his body was never found. General Kemper was shot from his horse and was crippled for life. General Armistead, who participated in the climatic assault against the Federal center, suffered a mortal wound at the Angle.

Brigadier General James Lawson Kemper (1823–95) *suffered a serious wound while leading his brigade in Pickett's unsuccessful assault on Cemetery Hill. Though desperately wounded, he survived and later became a major general and governor of Virginia.*

"I met Pickett's Division, returning after the battle," Hotchkiss had recorded in his journal, "scattered all along the road; no officers and all protesting that they had been completely cut up."

As the Confederates streamed back to Seminary Ridge, they observed Lee waiting to meet them. He looked pallid and downcast, and as the men passed, he said, "It's all my fault. My fault." Later, expecting trouble from Meade, Lee went to Pickett and instructed him to prepare to repulse a possible counterattack. Pickett replied, "General Lee, I have no division now." Lee later learned that Pickett had lost 3,000 men and all fifteen of his regimental commanders.

LEE RETREATS

Hotchkiss had watched the battle from the viewpoint of Seminary Ridge, and he observed the survivors of Pickett's division as they returned to the rear. "The unmistakable signs of retreat were plentiful," Hotchkiss wrote that night. "There was a general feeling of despondency in the army at our great losses, though the battle is regarded as a drawn one." Hotchkiss viewed the battle through rose-colored glasses because no one wanted to admit defeat. "I went back to the wagons," Hotchkiss added, "to work on a map." It was not a map of the battlefield but a sketch "of the country back to Virginia." Before dawn on July 4, Ewell's quartermaster, Major James Harman, began moving a wagon train of baggage and captured supplies to Williamsport. Hotchkiss eventually used the same map (see page 98, Map #158) to mark Ewell's advance into Pennsylvania as well as the lines of the Second Corps' retreat.

At 2:00 a.m. on July 4, Hotchkiss began sketching routes to get the shattered army quickly away from Gettysburg by using different roads. Ewell found a topographer in Harry Hays's brigade and sent Samuel B. Robinson of the Seventh Louisiana to help with maps. Hotchkiss quickly put Robinson, a skilled draftsman from England, to work on making copies. From that day forward, Robinson became a permanent member of Hotchkiss's topographical staff, as well as a lifelong friend, and later contributed several maps of future campaigns. In addition to being an excellent chef, Robinson filled a gap in Hotchkiss's life, which had been caused by the death of Boswell at Chancellorsville.

Lee expected to be attacked at any hour, but Meade resisted demands from Washington to pursue the enemy and bring the war to an end. Lee marveled at Meade's timidity. After heavy rain began to fall on the night of July 4, Lee ordered the withdrawal. The Federals followed without showing any enthusiasm, and even when the Potomac River flooded and prevented Lee from crossing, Meade made only halfhearted stabs at the beleaguered Confederates.

Hotchkiss went to Lee's headquarters every day for instructions. Until the Potomac receded, Lee wanted to establish fortified defensive positions and tasked Hotchkiss with the responsibility of finding them. While the army constructed eight miles of fortified earthworks on the north side of the Potomac, Hotchkiss roamed the area looking for fords but found none other than the one at Williamsport. On July 14, with the enemy finally pressing with a strong force, Lee's army crossed the river into Virginia, leaving the Federals behind. Four days later, Meade crossed the Potomac after squandering his best opportunity to shorten the war.

On July 30, most of the Confederates waded across the Rapidan and settled into defensive positions around Culpepper. While the army rested, Hotchkiss labored on his maps with help from Robinson and Brown. Generals wanted maps of Gettysburg for their after-action reports, but Lee was looking down the road to the next moves by the Army of the Potomac and kept Hotchkiss engaged in mapping the counties where he expected future battles to be fought.

On July 4, 1863, the skies opened up over the Gettysburg battlefield, and Lee's army quietly departed with wagons filled with the wounded. The lead wagon, and probably others, had been captured from the Federals and bore the inscription "U.S." on the wagon's canvas covering.

Brigadier General Lewis Addison Armistead (1817–63) *was one of three brigade commanders who lost his life during Pickett's Charge at Gettysburg. Armistead was the only general to reach the stone wall on Cemetery Hill before suffering a mortal wound.*

THE BATTLE OF BRISTOE STATION: OCTOBER 14, 1863

In late July, Meade moved the Army of the Potomac east of the Blue Ridge and began concentrating his forces along the Rappahannock River, opposite the Confederates. Lee had diluted his force by sending Longstreet with two divisions to Chickamauga, Georgia, in support of General Braxton Bragg. This gave Meade a two-to-one advantage, so Lee retired behind the Rapidan River. Lee's aggressive instincts began to percolate in early October when he learned that Union Secretary of War Stanton had detached the Eleventh and Twelfth Corps from the Army of the Potomac and sent them to Chattanooga, Tennessee, to reinforce General William S. Rosecrans.

Lee decided to worry Meade and drive him out of Virginia by assaulting the Federal right flank. He forced Meade to retreat but failed to prevent him from escaping to a defensive position near Centreville on the old Bull Run battlefield. Aside from minor skirmishes and wild cavalry clashes, the only important infantry engagement occurred at Bristoe Station, a few miles south of Manassas.

On October 14 A. P. Hill's Third Corps drew the assignment of flanking Meade's withdrawal from the Rappahannock. To get on Meade's flank, Hill marched hard and on the morning of October 14 reached Warrenton together with Ewell's Second Corps. Hill learned the long Union line had been moving along the tracks of the Orange and Alexandria Railroad and stretched for miles. Without performing routine reconnaissance, he hurried forward General Heth's division after observing Warren's Third Corps crossing Broad Run east of Bristoe Station.

Heth's two leading brigades deployed hurriedly on observing Hill seething with impatience. Without reconnoitering the position, Hill sent both brigades charging down the hillside and toward an obtuse angle formed by Broad Run and the railroad. Both brigades ran into a hail of fire from Federal infantry that had been carefully hidden and well placed to protect Meade's withdrawal. The Federals destroyed both Confederate brigades and wounded both brigade commanders. Hill demonstrated extremely poor generalship by failing to reconnoiter, and the episode marked his first defeat as a corps commander. Lee lost 1,900 men at Bristoe Station, the Federals 548.

After beating off a poorly prepared attack by the Federals at Mine Run in late November, Lee took the army into winter quarters and waited for spring. His officer corps had been decimated at Gettysburg, and the action at Bristoe Station only magnified the problem of finding competent field officers.

Disappointed by Hill's reckless generalship and somewhat embarrassed by the fiasco, Lee did not ask Hotchkiss for a map of the Bristoe Station affair until February. Instead, Hotchkiss and Robinson spent most of the winter months completing maps of the battle of Gettysburg, expanding the Shenandoah Valley map, building new maps of Virginia's counties, and satisfying requests for maps from Lee's commanders.

While cold winter days slowly passed and the army recovered, Hotchkiss never rested. His journal during those months read one day to the next, "worked hard all day and was very weary," or "spent the day on reconnaissance and returned chilled with map work to do." By March 1863, Hotchkiss had provided Ewell and Lee with maps of all the counties of Virginia from Richmond west to the Shenandoah Valley, and from Jefferson County on the Potomac to Lynchburg and Petersburg in southern Virginia.

Before the end of 1864, the Army of Northern Virginia would need them all.

During the battle of Bristoe Station, Confederate soldiers scattered. This illustration depicts one lone soldier who is still looking for his regiment.

Battle of Bristoe Station [Map #165]

The 1863 battle of Bristoe Station resulted from a series of maneuvers by General Lee in Prince William County to cut off Meade's withdrawal from the Rappahannock River while at the same time threatening Washington. Although the maneuvering occurred from October 9 to 22, the actual battle took place on October 14. Hotchkiss had been at home ill during the Bristoe Station affair and did not complete the map until February 21, 1864.

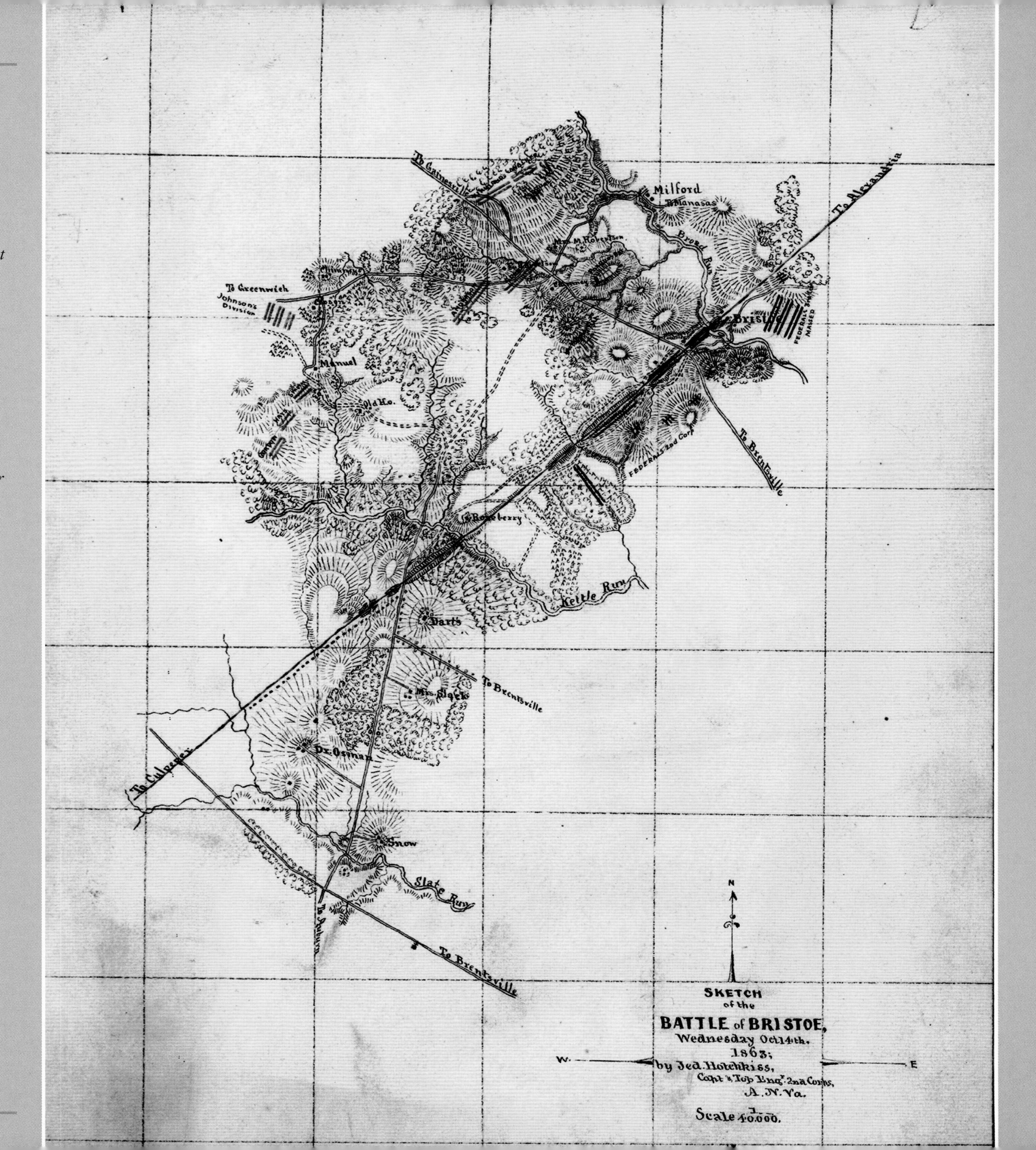

CHAPTER SIX: FROM THE WILDERNESS TO PETERSBURG

After the Union's Mine Run campaign sputtered on December 1, 1863, the Virginia landscape turned to mud and the armies went into winter quarters. Fifteen days later, Ewell sent Early's division to the Shenandoah Valley to quash Federal cavalry raids. On December 19 Hotchkiss obtained a short leave and went home. Ewell decided that his topographer could get more work done if he had better accommodations, so Hotchkiss rented an office in Staunton, brought Robinson from camp, and hired C. William Oltmanns, a local draftsman. The two men—Robinson and Oltmanns—remained with Hotchkiss for the remainder of the war. The rented office became Hotchkiss's headquarters and he used it except when in camp or in the field.

Wintering in comfort in Staunton gave Hotchkiss little relief from work. According to his journal, a day rarely passed without working on a battle map, a county map, or responding to a request for information from one of the generals. The work progressed rapidly because Robinson and Oltmanns did the copying and Hotchkiss did the surveying and scouting. As all the generals used the winter months to file battle reports, Colonel William Proctor, Lee's chief of engineers, asked for maps of Bristoe Station, Mine Run, Culpepper County, a reconnaissance in Fauquier County, maps of Pennsylvania, maps of Maryland, and maps of the Richmond area.

On January 21 Hotchkiss received new orders from General Early, who commanded the Valley District. The instructions probably came from Lee, who was anticipating the spring campaign and wanted the mountains south of Staunton and west of Lexington scouted and surveyed for defensive purposes. Hotchkiss had never mapped the region for Jackson, so he spent two months in the saddle rambling through Rockbridge County with a trooper from the cavalry. They rode together many miles each day in rain, snow, sleet, and wind during one of the most brutal winters in the history of the upper valley. Hotchkiss made frequent stops at the Staunton headquarters in order to keep Oltmanns and Robinson working on a steady inflow of new projects.

On April 18 Hotchkiss returned to army headquarters at Orange Court House with the map of Rockbridge County and became immediately aware of apprehension in camp caused by an increase in Federal activity across the Rapidan River. "Everything indicates," he wrote on May 1, "that in less than 48 hours we may have the bloodiest battle of the war fought on this very ground."

THE GATHERING OF ARMIES

Since the outbreak of the Civil War in April 1861, President Lincoln had been experimenting with generals, searching to find someone to run the Union field operations. Until forty-one-year-old Ulysses S. Grant appeared on the scene, fresh from victories at Vicksburg in July and Chattanooga in November, Lincoln had not been able to find an acceptable fighting general capable of taking the load of military management from his shoulders. Grant may not have been the perfect candidate for assuming the role of commander in chief of the Union armies, but from Lincoln's perspective Grant was the only candidate. Lincoln had tried McClellan, Burnside, Hooker, and Meade as field commanders of the Army of the Potomac, and in the president's opinion, they had all wasted opportunities to shorten the war. When Grant decided to establish his command headquarters with the Army of the Potomac instead of in Washington, Lincoln expected the general to remove Meade. Grant surprised Lincoln. He retained Meade as the Army of the Potomac's commander, but every campaign that followed was the creation of Grant's strategy and not Meade's. He simply told Meade, "Lee's army will be your objective point. Wherever Lee goes, there you will go also."

Grant now had 662,000 soldiers spread between twenty-two corps with which he could do whatever he wanted so long as he protected Washington. Meade had rebuilt the Army of the Potomac during the winter, and on May 1, 1864, it contained about 120,000 infantry, artillery, and cavalry. Lee had 60,000 men, about 20,000 fewer than he had taken to Gettysburg.

Lee expected Grant to fight and make every effort to force the Confederates into the open. With such a numerical disadvantage, Lee could no longer risk fighting major offensive battles. On May 2 he rode to Clark Mountain to personally reconnoiter the Federal position, which spread for miles along the opposite bank of the Rapidan. From the crest, he looked down on the muddy, 200-foot-wide river and could see Ely's Ford a few miles upriver and Germanna Ford thirteen miles downriver. South of the fords spread the vast and tangled Wilderness,

ROBERT EDWARD LEE (1807–70)

Born in Westmoreland County, Virginia, Lee was the fifth child of Henry "Light-Horse Harry" Lee of Revolutionary War fame. Henry died when Lee was eleven years old. After an early education in Alexandria, Virginia, Lee obtained an appointment to West Point and in 1829 graduated second in a class of forty-six and joined the Corps of Engineers. In 1831 he married Mary Ann Randolph Custis, the great-granddaughter of Martha Washington. During the Mexican War, Lee suffered one wound, earned three brevets for gallantry, and emerged with a reputation for brilliance on the battlefield.

On April 18, 1861, and with the venerated General Winfield Scott's recommendation, President Lincoln offered Lee command of the Union army. Lee declined and two days later took command of Virginia's troops. He accomplished little as a military adviser to Jefferson Davis or when campaigning in western Virginia, but on June 1, 1862, after General Joseph Johnston suffered a wound during the Peninsula campaign, Lee took the reins of the Army of Northern Virginia.

For nearly three years, with inferior numbers and few resources, Lee faced the Union's powerful Army of the Potomac and at one time or another embarrassed every general who commanded that army. He out-generaled McClellan during the Seven Days campaign, defeated Pope's army at Second Manassas, fought McClellan to a draw at Antietam, bloodied Burnside's army at Fredericksburg, and in perhaps the most remarkable campaign of the entire war, outflanked and defeated Hooker at Chancellorsville. In the spring of 1864 he faced a new opponent, Ulysses S. Grant, a man with all the resources of the North and a force twice the size of the Army of Northern Virginia. Grant began pushing Lee around in the Wilderness campaign in what ultimately became an eleven-month war of attrition, a contest the South could not win. Lee used all his resources to frustrate Grant. One of those resources was knowledge of the terrain. Jed Hotchkiss commanded a small topographical staff that created and reproduced maps for Lee and his generals. By 1863 Hotchkiss had become the topographical clearinghouse for the Army of Northern Virginia.

After the war, Lee returned to Richmond as a paroled prisoner and devoted his life to rebuilding Washington College in Lexington, Virginia. With Lee involved, young Northerners enrolled right along with Southerners. Lee died in 1870 after making the college among the best in the South. For reasons that require no explanation, it is known today as Washington and Lee University.

some twelve miles wide and six miles deep, choked with a mass of oak, pine, and scrub brush. Having studied Grant's options, Lee put down his glasses and said, "Grant will cross by one of those fords." Lee's ally would not be numbers—it would be the Wilderness.

THE WILDERNESS CAMPAIGN: MAY 5, 1864

Shortly after midnight on May 4, 1864, General Warren's Fifth Corps and General Sedgwick's Sixth Corps began crossing the Rapidan on pontoons at Germanna Ford. Once across, the long column formed on the Germanna Plank Road and marched toward Wilderness Tavern on the Old Turnpike. General Hancock's Second Corps crossed simultaneously using Ely's Ford. General Burnside's Ninth Corps followed next, using Germanna Ford. Grant intended to turn the Confederate right flank and press the enemy toward Orange Court House. He did not expect to fight in the Wilderness.

About 9:00 a.m. on May 4, Lee learned from Stuart's scouts that Federal forces had crossed the Rapidan. He received the news at his headquarters at Orange Court House, twenty miles away. He immediately informed Ewell at Mine Run, thirteen miles south of Wilderness Tavern, and Hill, whose corps lay another seven miles south of Ewell. Thirty miles away, Longstreet's First Corps were on a foraging expedition when Stuart's messengers arrived. Ewell immediately started the divisions of Early, Johnson, and Rodes toward the Old Turnpike to prevent the Federals from interposing between the three Confederate corps or advancing westward on the Old Turnpike toward Lee's headquarters at Orange Court House. Ewell intended to keep Hotchkiss away from the fighting, leaving him behind "to receive and give orders" should any be sent from Lee. While Ewell moved from Mine Run, Hill began marching toward the Old Turnpike using the Orange Plank Road. All the roads in the area connected with either the Old Turnpike or the Orange Plank Road, both of which ran through the Wilderness and became central to the evolving battle. When Ewell's corps halted and bivouacked on the night of May 4 at Locust Grove on the Old Turnpike, his corps and Warren's Federals were five miles apart. Neither knew the location of the other.

On the morning of May 5, and following Lee's orders to engage any force encountered, Ewell moved eastward along the turnpike and into the Wilderness. Instead of remaining behind, Hotchkiss scouted ahead of the Second Corps' right flank and carried messages between Ewell and Lee. At the same hour, Warren's blueclads began moving west on the turnpike because Grant wanted to get Federal forces out of the Wilderness. At 7:00 a.m., Ewell's and Warren's skirmishers collided about two miles west of Wilderness Tavern. Grant wanted action, so Meade instructed Warren to attack immediately with his entire force. Meade also ordered Sedgwick, whose Sixth Corps were still on the Germanna Plank Road, to advance in support of Warren's right, and for Burnside to bring the Ninth Corps immediately across the ford to support Sedgwick.

Ewell's corps flared into the woods with Early on the left, Johnson in the center and straddling the turnpike, and Rodes on the right. Brigadier General Charles Griffin's Federal division had hurriedly thrown up breastworks. His brigades then pressed forward to determine the size of the approaching Confederate force. Brigadier General James Wadsworth's division supported Griffin's left, and Sedgwick sent Brigadier General Horatio G. Wright's division down a near-impassable Wilderness road clogged with overgrowth in which the Federals became entangled until late in the afternoon.

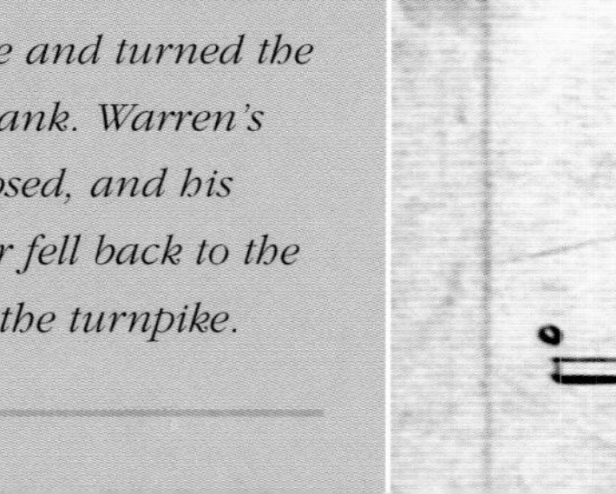

Battle of the Wilderness: The First Day [Sketchbook Map #8-2A]
The first day's engagement in the Wilderness on May 5, 1864, pitted Edward Johnson's division and Robert Rodes's division against General Warren's Fifth Corps. Both forces were aligned in the woods on both sides of the Old Turnpike (Orange–Fredericksburg Turnpike). Hotchkiss's map shows the position of forces without showing the action. At the beginning of the fight, Federal forces broke through the center of the Confederate line. Early's division rushed into the melee and turned the Federal left flank. Warren's assault collapsed, and his divisions later fell back to the north side of the turnpike.

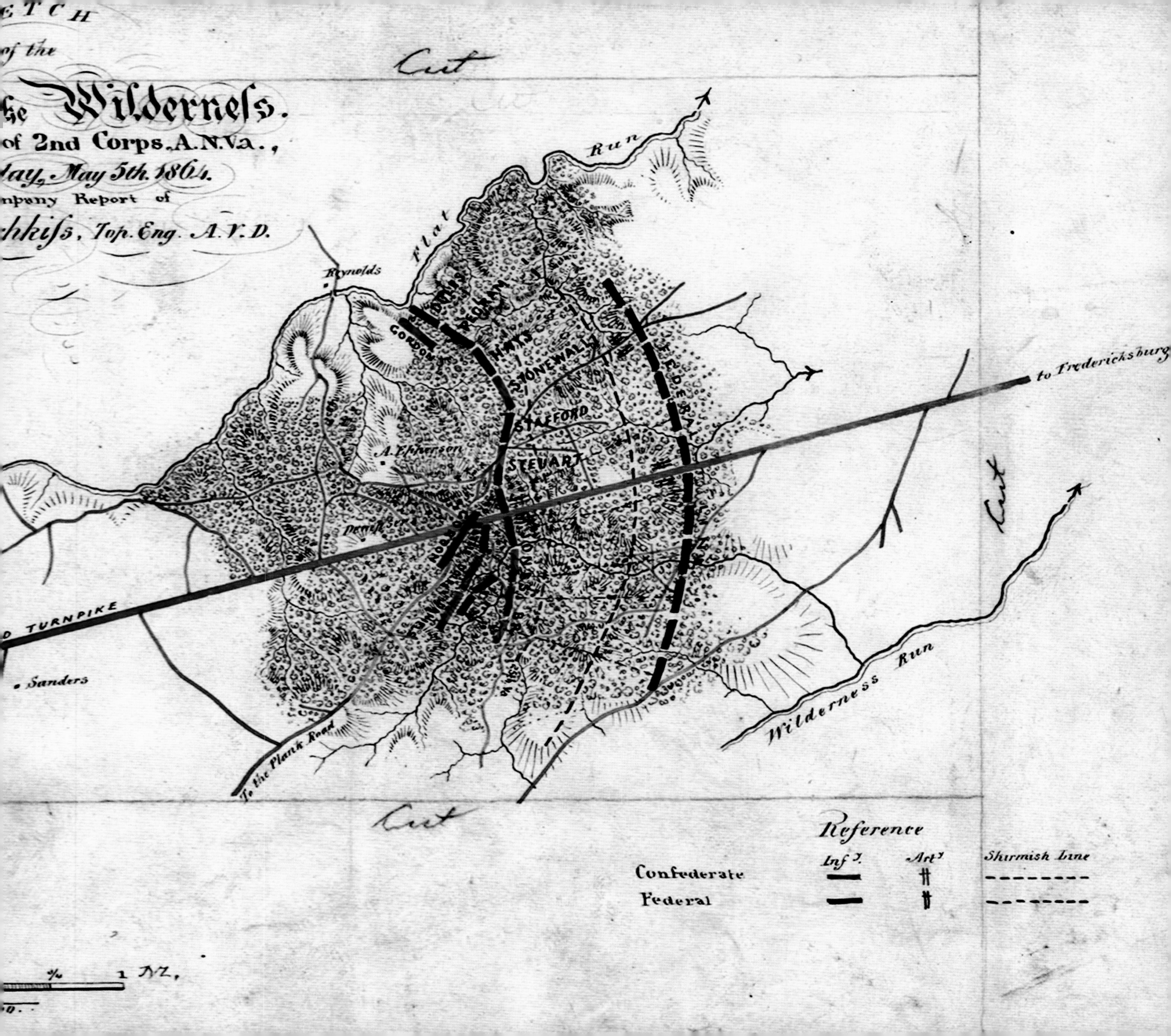

etch
of the
he Wilderness.
of 2nd Corps, A. N. Va.,
day, May 5th. 1864.
mpany Report of
chkiss, Top. Eng. A. V. D.
Cut
Flat Run
Reynolds
GORDON
PEGRAM
HAYS
STONEWALL
STAFFORD
STEUART
A. Epperson
FEDERAL
to Fredericksburg
TURNPIKE
Sanders
To the Plank Road
Wilderness Run
Reference
Confederate
Federal
Inf.y
Art.y
Skirmish Line
1 M.

As Lee approached, he became aware that Ewell was pitted against a major force, and at 8:00 a.m. suggested that the latter moderate his advance to allow Hill more time to work up the Orange Plank Road and get into the enemy's rear. Lee had not yet observed the approach of Hancock's Second Corps on the Brock Road, which lay six miles southeast of Wilderness Tavern. Lee's order reached Ewell too late. By 9:00 a.m., both Ewell's and Warren's corps were dug in and fighting furiously in the Wilderness. Flash fires erupted, smoke poured through the woods, and at 1:00 p.m. a major engagement developed that raged through an afternoon of assaults and counterassaults. Wounded men on both sides perished in the fires that engulfed sections of the Wilderness. Ewell's corps finally repulsed Federal advances, including Sedgwick's attempt to strike the Confederate left flank, and the firing gradually simmered to a halt at nightfall.

At 11:30 a.m., Henry Heth's division from Hill's Third Corps, after being annoyed all morning by Federal cavalry, collided with Brigadier General Richard Getty's Federal brigade from the Sixth Corps near the intersection of the Orange Plank Road and the Brock Road. Lee wanted control of the intersection, and a second fight erupted. Behind Heth came the rest of Hill's corps, but Hancock's Second Corps were also fast approaching the intersection from the Brock Road with another column of blueclads. At 4:00 p.m., Hancock's brigades joined Getty and charged. The Federal units performed poorly and the attack failed, mostly because the blueclads encountered difficulty forming in the Wilderness. Federal officers became disoriented, and the planned coordinated assault disintegrated into separate, disjointed attacks that failed altogether. Darkness intervened, which had been the hopes and prayers of men on both sides, and the fighting subsided.

THE WILDERNESS CAMPAIGN: MAY 6, 1864

During the night, General Grant learned from captured prisoners that Longstreet had not yet joined Lee. He ordered a 5:00 a.m. assault by Sedgwick's Sixth Corps and Warren's Fifth Corps against Ewell on the Confederate left while Hancock's Second Corps assaulted Hill on the Confederate right. Burnside's Ninth Corps filled the large gap between Ewell and Hill for the purpose of supporting either Hancock or Warren. Lee issued orders for Ewell to open up with artillery early on May 6 to take pressure off Hill. At 4:30 a.m., Ewell attacked Sedgwick's Sixth Corps but was driven back by Federals about to launch their own assault. Ewell's grayclads withdrew to defensive positions and checked advances by Sedgwick and Warren throughout the entire day.

Ewell's early morning assault did not slow down Hancock. At 5:00 a.m., 20,000 blueclads stretched out in three lines of battle more than a mile wide. They rambled into the Wilderness and began rolling over Hill's skirmishers. Burnside arrived late. Instead of immediately striking Hill's left flank, he delayed and allowed Ewell to move Rodes's division into the gap. With his advance checked and stymied by the confusing terrain, Burnside did nothing.

As soon as the Federals stepped off the Brock Road and into the Wilderness, Hill's brigades wilted under the weight of the attack, and grayclads began running to the rear. Brigadier General James A. Wadsworth's division from Warren's Fifth Corps began enveloping Hill's left flank while Major General David Birney's division enveloped Hill's right flank. Hancock began capitalizing on what appeared to be a Confederate rout and plunged his divisions more than a mile into the Wilderness. Federal brigades and regiments became hopelessly

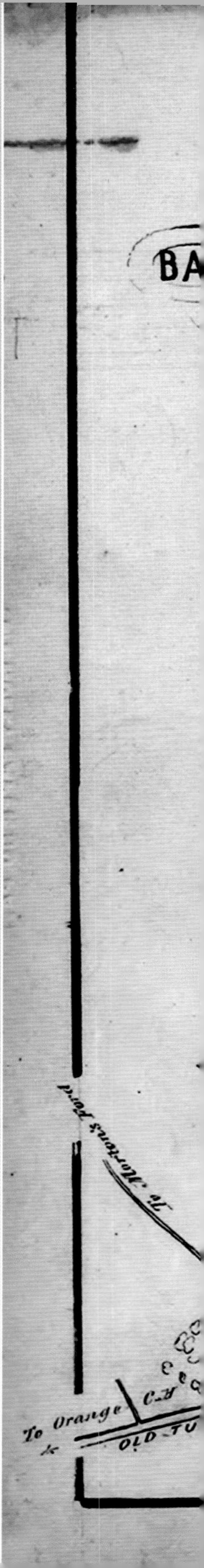

Battle of the Wilderness: The Second Day [Sketchbook Map #8-3]
The early action in the Wilderness on May 6 involved four Federal corps pitted against Lee's two Confederate corps. Longstreet had not yet arrived. The engagement occurred roughly in the same two areas where it had ended the night before. The Hotchkiss map shows only the position of Ewell's Second Corps and not the position of Hill's corps on and around the Orange Plank Road, where the major engagement of the day occurred.

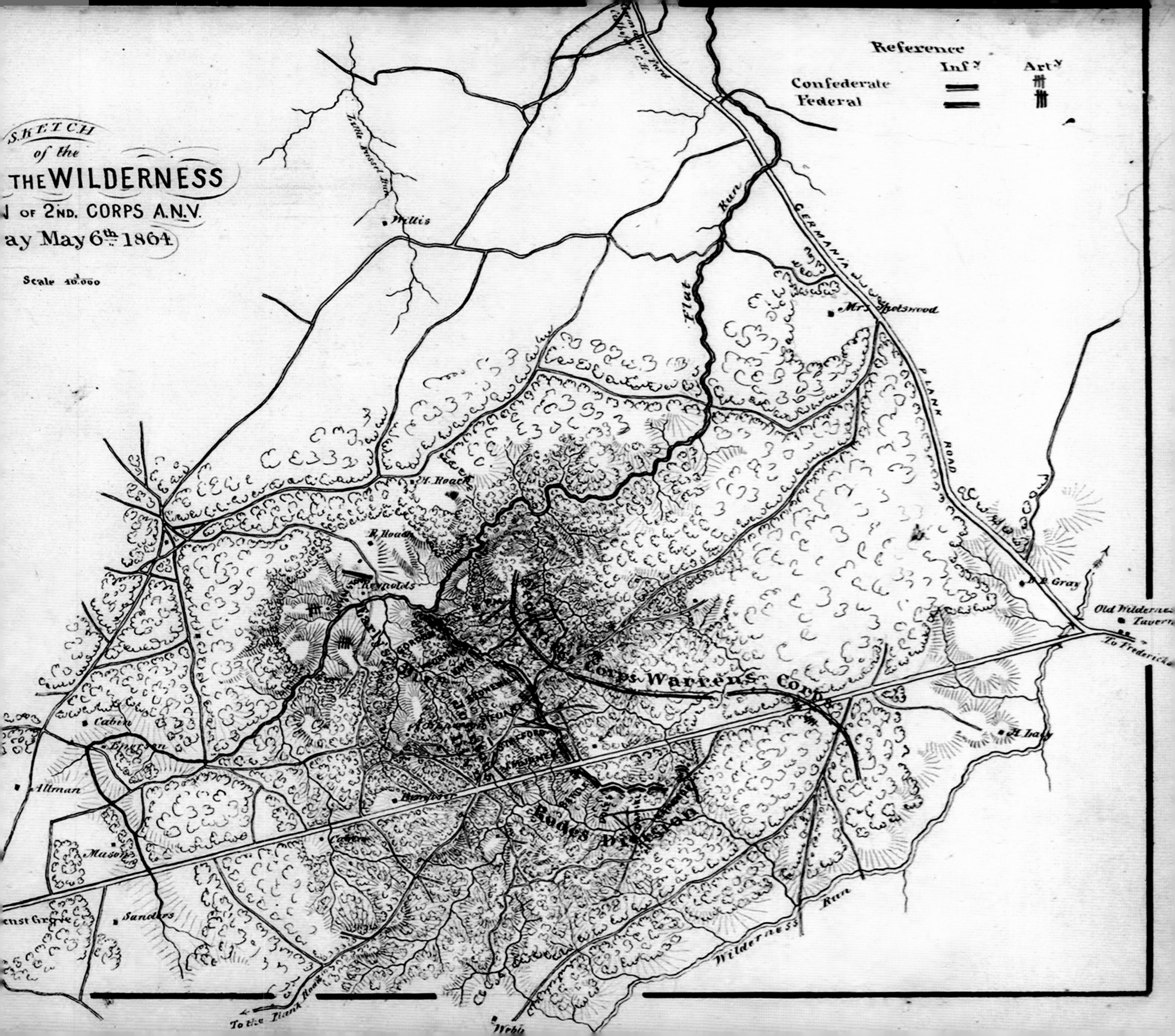

SKETCH
of the
THE WILDERNESS
OF 2ND. CORPS A.N.V.
ay May 6th 1864
Scale 1/40.000
Reference
Infy
Arty
Confederate
Federal
Little Wilderness Run
Willis
Flat Run
GERMANIA PLANK ROAD
Mrs Spotswood
M. Roach
Reynolds
Corps Warrens Corps
Old Wilderness Tavern
To Fredericks
Cabin
Emerson
Altman
Mason
Sanders
Rodes Division
Wilderness Run
To the Plank Road
Webb

entangled, firing blindly into the smoke and often at their own men. Dry vegetation caught fire from the discharge of muskets, and when Confederate artillery began to hammer the woods, larger fires erupted. The disoriented blueclads lost their sense of direction and scattered throughout the burning brush.

Lieutenant General Ulysses S. Grant (1822–85) *became general-in-chief of Federal forces on March 9, 1864. He made his headquarters with the Army of the Potomac and took over the strategic direction of the war. Although Grant racked up severe casualties while ending the war, in 1869 he became the eighteenth president of the United States.*

> *"All circumstances seemed to combine to make the scene one of unutterable horror. It was as though Christian men had turned to fiends, and hell itself has usurped the place of earth."*
>
> —LIEUTENANT COLONEL HORACE PORTER, U.S. ARMY

Hancock's forces had come within reach of a glorious victory when Kershaw's division from Longstreet's corps, marching in parallel columns on the Orange Plank Road with Major General Charles W. Field's division, pressed through the wreckage of Hill's corps and stopped the Federals in their tracks. Longstreet came on the field, reorganized Hill's scattered brigades, checked Hancock's advance, and closed the gap between Ewell and Hill. Longstreet counterattacked across the Federal front, sent Wadsworth's blueclads fleeing in different directions, and pushed Hancock's commingled brigades one upon another. Longstreet reestablished a strong defensive perimeter, and further efforts to penetrate the Confederate line by Hancock failed.

By 11:00 a.m., the fighting began to show signs of abating, but Longstreet still had cards to play. Hotchkiss had been carrying orders and making sketches of the battlefield as he moved between Ewell's corps on the left, Lee's field headquarters in the rear, and Longstreet's corps on the right. At one point, Hotchkiss narrowly escaped death when a minié ball struck and destroyed the binoculars that hung from his neck over his heart. He probably understood the network of roads and logging trails that crossed through the Wilderness better than anyone. He had mapped the Wilderness prior to the Chancellorsville campaign and had provided Jeb Stuart and General Jackson with sketches of the entire area. During the morning's battle, Stuart located a shielded approach through back roads that led to Hancock's left flank. Lieutenant Colonel Moxley Sorrel, Longstreet's adjutant general, followed a route traced on a Hotchkiss map and with four brigades rolled up Hancock's flank. Longstreet simultaneously attacked from the woods and drove Hancock's forces back to the Brock Road.

Stuart located another back road that led directly to Hancock's repositioned left flank on the Brock Road. As Longstreet formed concealed units for the assault, Sorrel's returning brigades mistakenly fired into the formation. The racket alerted the enemy, and Longstreet, having suffered a wound from friendly fire, aborted the assault, turned the corps over to Dick Anderson, and was carried to the rear. Lee came on the field and, as usual, suspected the Federals could be whipped. At 4:00 p.m., he launched a massive frontal attack against Hancock, followed a short time later by Ewell's assault on Sedgwick's flank. Lee's delay in organizing the assault gave the Federals time to strengthen a defensive position backed by artillery. Both assaults failed, and the battle of the Wilderness ended at 6:00 p.m.

There are no exact numbers of casualties, only estimates. Lee's army contained about 60,000 effectives and reported losses of about 7,750 killed, wounded, or missing. Meade's army contained from 115,000 to 120,000 effectives and sustained about 17,666 casualties, which included more than 3,000 missing or captured. Many of the bodies lost in the Wilderness were burned beyond recognition. "We buried about 1,500 Yankees today," wrote Hotchkiss on May 7, "and collected many arms."

THE ROAD SOUTH

Hotchkiss expected the usual reprieve after a heavy battle and put Robinson and Oltmanns to work on maps of Richmond, which Ewell wanted as anxiously as Lee. But on May 7, while Confederate infantry buried the Federal dead, Grant did the unexpected. Following every battle in the past, the Army of the Potomac returned to a strong defensive position to regroup after suffering heavy casualties. The Federals did not attack on May 7 but sent a long train of ambulances toward Fredericksburg, and this alerted Lee to the possibility that Grant intended to move either forward or sideways. Grant had no intention of retreating, but he did intend to steal a march on Lee. Grant looked to Spotsylvania because its roads led to Hanover Junction, Lee's principal supply base. He also intended to interpose the Federal army between Richmond and the Army of Northern Virginia and force Lee into a running battle.

Wary of Grant's behavior, Lee ordered a track cut through the Wilderness to shorten the distance between the Orange Plank Road and the Carpathian Road, which led directly to Spotsylvania. At 8:30 p.m. on May 7, the race to Spotsylvania began when Grant sent Meade's army down the Brock Road, passing behind Hancock's corps, which blocked the Orange Plank Road. When Hill reported Federal guns moving down the Brock Road, Lee took no chances of being flanked and at 11:00 p.m. put Longstreet's corps on the road toward Spotsylvania, with Anderson in command.

THE BATTLE OF SPOTSYLVANIA: MAY 8–19, 1864

At dawn on May 8, Anderson stopped three miles from Spotsylvania to rest his men and allow them to prepare breakfast after an arduous and sleepless night march. Federal cavalry were already dueling with Stuart, who sent messengers to Anderson asking for infantry support at Spotsylvania Court House. Anderson put two weary brigades on the road. When Federal cavalry observed Confederate infantry approaching, they abandoned Spotsylvania and informed Grant that Longstreet's corps had arrived. In the meantime, A. P. Hill requested sick leave, and Jubal Early assumed temporary command of the Third Corps.

Hotchkiss was always trying to deflect his thoughts from the savagery and bloodletting of the battlefield, calming his thoughts with pleasant distractions. After witnessing new horrors in the Wilderness, he rode into Spotsylvania, jotting in his journal, "The trees are just getting fully out; the apples and peaches are in full bloom." In the hundreds of pages in his journal, and in the huge collection of his letters home, he never spoke hatefully of the enemy.

By the morning of May 9, Confederate forces had entrenched, blocking Grant's line of advance. Because Burnside, who had taken possession of the road to Fredericksburg, had misinformed Grant of Lee's dispositions, Meade ordered three divisions from Hancock's Second Corps to attack across

Lieutenant General Richard Heron Anderson (1821–79) *served as one of General Lee's most trusted generals. Although he fought as a division commander in Longstreet's corps, he led the corps whenever the latter was on detached duty or wounded.*

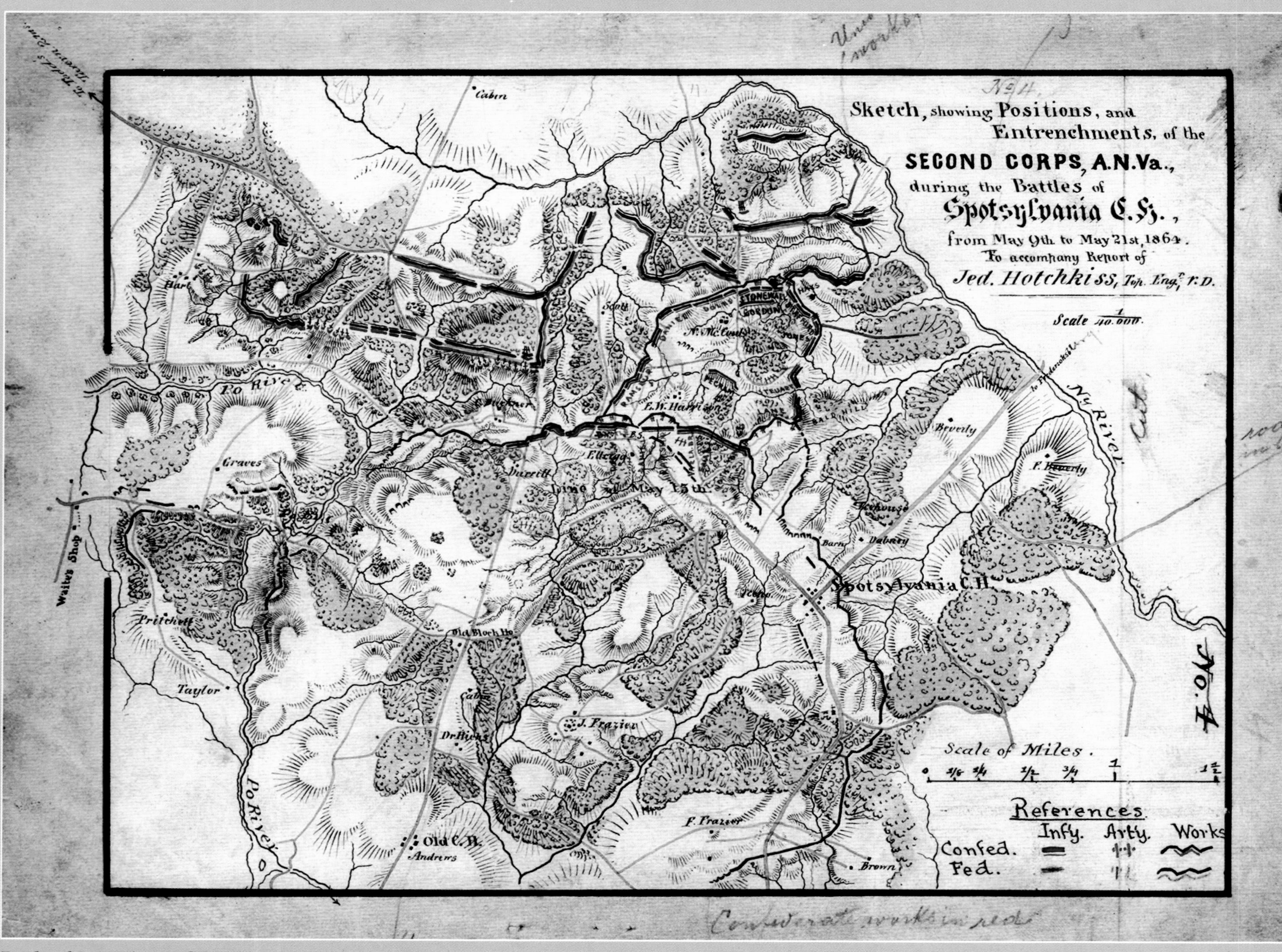

Battle of Spotsylvania [Sketchbook Map #8-4]

Hotchkiss's map shows the general arrangement of Confederate and Federal forces as a major battle erupted on May 10. The section to the right of the Po River became the focus of the Federal assault on May 9–10. The salient at the top of the Confederate line, called the "Mule Shoe" and "Bloody Angle," became the focus of the Federal assault on May 12. Grant lashed at other sections of the Confederate line at the same time, and the tide of battle ebbed and flowed for twelve days.

Major General Harry Thompson Hays (1819–75) *served under Jackson, Ewell, and Early in most of the campaigns of the Army of Northern Virginia until wounded at Spotsylvania, after which he recovered and finished the war in the Trans-Mississippi theater.*

the Po River and strike Lee's reportedly unguarded left flank. Brigadier General William "Little Billy" Mahone's division waited in a strong defensive position and checked the Federal assault.

Lee became concerned throughout the day that Grant's forces would flank the Confederates and regain the Brock Road, which passed through Spotsylvania Court House to an intersection that connected with the Telegraph Road to Richmond. Hotchkiss witnessed very little of the battle that day and worked until dusk copying maps of the area around the Confederate capital and the roads leading south from Spotsylvania.

On the morning of May 10, Heth's division moved around Mahone's left and assaulted Hancock's right flank. The Federal brigades reeled back and retired to the rear. Thinking that Lee had weakened his lines to reinforce Mahone, Grant sent Warren's Fifth Corps and Wright's Sixth Corps against Longstreet's First Corps, commanded by Anderson. Due to the lateness of the hour and confusion in organizing the attack, the poorly coordinated Federal assault collapsed and fell back with heavy losses. "I worked at maps, in the field," wrote Hotchkiss, noting that during the afternoon, "Early moved to our left and drove the enemy back across the [Po] river."

THE "MULE SHOE" AND "BLOODY ANGLE"

The other Federal attack on May 10 occurred at the "Mule Shoe," so named because of its shape. Defensive works were usually arranged in a line of trenches with abatis in front and with batteries perched behind the infantry and on higher ground. Instead, Lee's engineers constructed a salient measuring a mile deep and half a mile wide, with a toe in the shape of an angle at the tip. As the Mule Shoe

Brigadier General George Hume "Maryland" Steuart (1828–1903) *was one of the few generals from Maryland serving the Confederacy. He commanded a brigade at Gettysburg, the Wilderness, and at Spotsylvania, where he was wounded at the Bloody Angle.*

occupied the very center of the Confederate line, it naturally attracted the attention of the Federals. From Ewell's corps, Gordon's division defended the right side of the shoe and Johnson the left. Ewell placed Rodes's division near the base of the salient to protect the flanks and provide reinforcements.

Brigadier General James Alexander Walker (1832–1901) *commanded the Stonewall Brigade at Gettysburg, Bristoe Station, the Wilderness, and at Spotsylvania, where he suffered a serious wound.*

At the crack of dawn on May 10, twelve Federal regiments arranged in four compact lines charged the left (west) flank of the Mule Shoe and penetrated the Confederate line. A second Federal division failed to follow as planned. Gordon's division counterattacked and repulsed the assault. Grant observed that the Mule Shoe could be breached and made plans to try again.

Although May 11 remained a relatively uneventful day at Spotsylvania, Lee soon discovered that Jeb Stuart had been killed at Yellow Tavern, a small village at a crossroad leading to Richmond. On May 9, Grant had detached Major General Philip H. Sheridan's cavalry to raid Lee's communications and threaten Richmond. Stuart was shot from his horse and mortally wounded while trying to repel the attack.

At 4:30 a.m. on May 12, Grant sent 20,000 blueclads in four divisions from Hancock's Second Corps against the apex of the Mule Shoe in a determined effort to break through the center of the Confederate line and force Lee into the open. After heavy fighting on May 10, Ewell had moved Gordon's division from the left of the Mule Shoe to the base. Rodes's division now occupied the left while Johnson defended the right. Within fifteen minutes after assaulting Johnson's division at the tip of the Mule Shoe, Hancock's blueclads opened gaps and poured into the salient. A furious fight ensued in an action known as the "Bloody Angle," which began at the apex of the Mule Shoe. Despite a heavy rainstorm, the battle lasted twenty-four hours in some of the most vicious hand-to-hand fighting of the war. At one point, an oak tree twenty-two inches in diameter toppled to the ground after being cut through by minié balls. So many bullets struck some men that their bodies fell apart. When General Lee observed the grayclads falling back, he came on the field to lead a counterattack. Grant almost got his wish, but grayclads gathered around Lee shouting "No, no!" and "General Lee to the rear!" Gordon hastily mustered his division and led a counterattack, which almost succeeded in driving the Federals out of the Mule Shoe when Wright's Sixth Corps struck Rodes's division and opened a huge breach through which thousands of blueclads poured.

The assault virtually destroyed the venerated Stonewall Brigade and wiped out most of Allegheny Johnson's division. By nightfall, twenty-four Federal brigades had plunged into the Mule Shoe. Artillery fire piled up Federal bodies three and four deep. In a fifteen-by-twelve-foot area of the Bloody Angle, 150 bodies were later discovered in a single pile. In less than a square mile, 12,000 men died.

Ewell's officer's corps lost heavily that day. The Federals captured Johnson, one of Ewell's most dependable division commanders, and wounded several brigade commanders, including "Maryland" Steuart, Harry Hays, and James A. Walker, all veteran officers who had been with the Second and Third Corps since their formation. Ewell eventually restored a defensive line at the base of the salient with help from Kershaw's and Wilcox's divisions, and at midnight the fighting finally ceased. Blasted by artillery and ripped apart by charging blueclads, the Mule Shoe and its Bloody Angle no longer existed. The Confederates erected a new defensive line at the base of the former salient, thus foiling Grant's efforts to force the fight into the open.

Throughout the day, Hotchkiss found himself much closer to the fighting than usual. "We lost many valuable lives," he wrote that night. "I spent the day, with many of the staff, rallying and sending in again the broken fragments of Johnson's Division and giving orders . . . in the vicinity of the battleground." Though always reluctant to risk his topographer on the front lines, Ewell needed every man at the Mule Shoe.

> *"The world has never seen so bloody and so protracted a battle as the one being fought and I hope never will again."*
>
> —ULYSSES S. GRANT TO HIS WIFE, MAY 13, 1864

EDWARD JOHNSON (1816–73)

Born is in Salisbury, Virginia, Johnson graduated from West Point in 1838, thirty-second in a class of forty-five. He served during the Seminole War and earned two brevets during the Mexican War. On June 10, 1861, he resigned as a captain of U.S. Army regulars in New York and became colonel of the Twelfth Georgia Infantry Regiment.

Johnson, Longstreet, Lee, Ewell, and Pickett were all personal friends from the Mexican War. Because of Johnson's well-known fighting qualities, Lee elevated him to brigadier general on December 13, 1861, to take command of a brigade in western Virginia. In a war where almost everyone had a nickname, "Old Allegheny" and "Allegheny Ed" seemed to fit Johnson because most of his early service occurred in the Allegheny Mountains. It was there that he became attached to Stonewall Jackson's division after the Battle of McDowell, and it was there that he suffered a shot through his foot.

Johnson was a man of great warmth, which appealed to Hotchkiss, and a man of many peculiarities, which Hotchkiss found amusing. They began to work closely during the Shenandoah Valley campaign of 1862. Hotchkiss had many conversations with Johnson, sometimes because General Jackson wanted assurance that Johnson understood the orders and sent Hotchkiss to go over maps. During a friendly conversation with Hotchkiss on July 3, 1862, Johnson said, "Too bad to be wounded after the first little fight [at McDowell]; I would not have cared after some of the big battles."

Promoted to major general on February 28, 1863, Johnson took command of the Stonewall Division after Jackson's death at Chancellorsville. He led the division at Gettysburg, and later, when conferring with Hotchkiss on a map, confessed that he "never wanted to go up the hill at Gettysburg," meaning Lee's poorly conceived Confederate assault on Culp's Hill.

Johnson continued to lead the Stonewall Division at the Wilderness and Spotsylvania. Caught in the chaos at the Bloody Angle when Federals overran the position, Johnson became the prisoner of a group of blueclads. Because the wound he received at McDowell healed poorly, Johnson had adopted the habit of carrying a long hickory cane into battle. He knocked a few Yankee heads with it before an old friend in the Union army, Major General Marsena R. Patrick, recognized Johnson, observed his irritation at being captured, and rescued him from further damage.

Eventually released, Johnson commanded a division during the Franklin and Nashville campaign in December 1864, only to be captured again and not released until July 1865. John E. Stanchak, who contributed to Patricia L. Faust's *Encyclopedia of the Civil War*, wrote that Johnson "was remembered as a gruff, badly dressed, voluble swearer, and as one of the Confederacy's best division commanders."

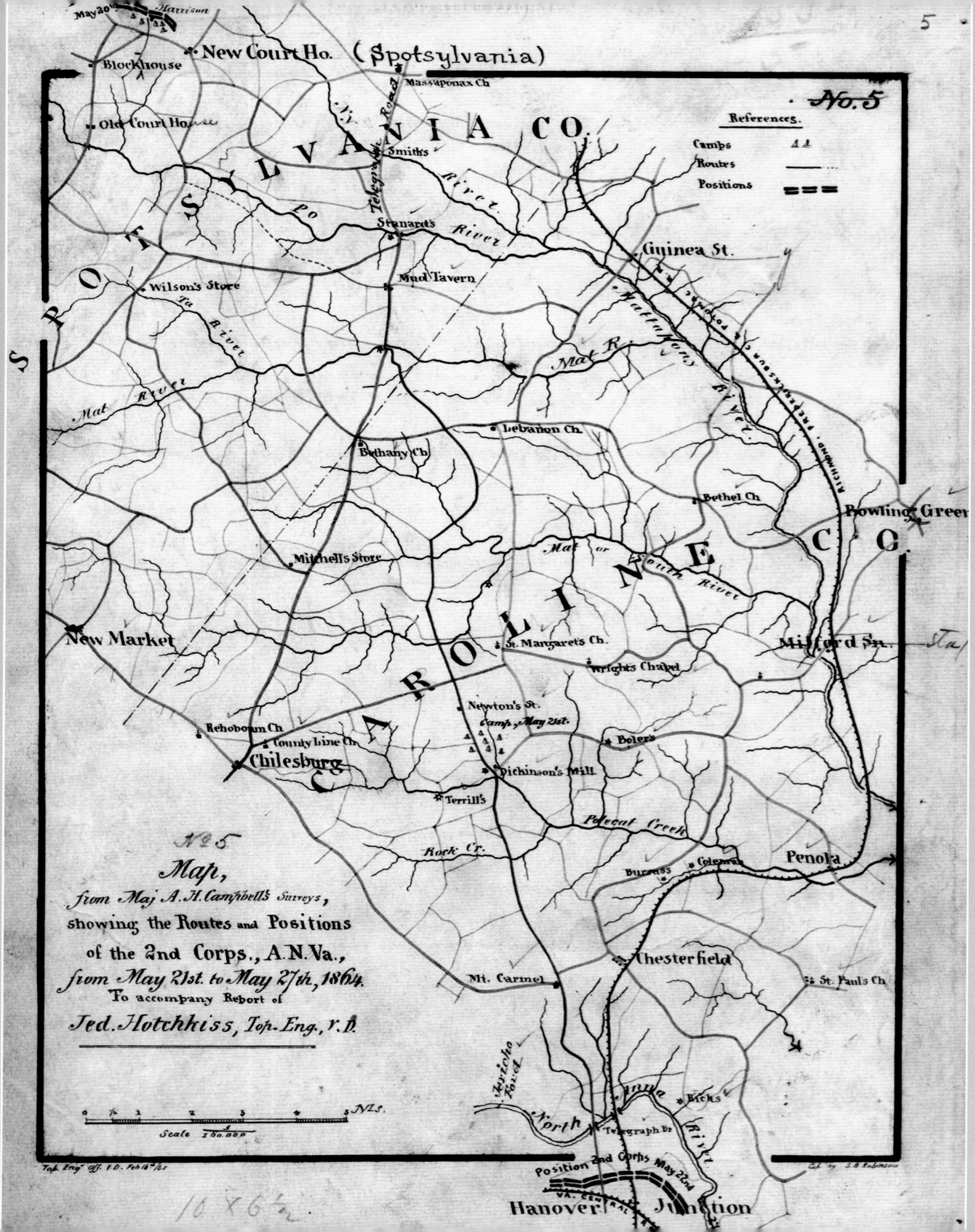

Caroline County
[Sketchbook Map #8-5A]

Hotchkiss built many maps of Virginia's counties because General Lee and General Jackson wanted foreknowledge of roads, streams, bridges, fords, and terrain. General Jackson had ordered a map of Caroline County on December 4, 1862, a few days before the battle of Fredericksburg. Once Hotchkiss started a map, he never stopped improving it. Here he connected parts of lower Spotsylvania County with upper Caroline County to map the quickest routes from Spotsylvania Court House to Hanover Junction. As a result of damage to Spotsylvania Court House in 1864, the rebuilt village went through a series of name changes, including New Court House, New Spotsylvania, and finally just Spotsylvania.

THE SPOTSYLVANIA FINALE

Throughout the Spotsylvania engagement, Grant continued to receive reinforcements while Lee fought with the ragged grayclads he had brought from the Wilderness. For several days, Grant moved his corps from one end of the Confederate line to the other in an effort to breach the enemy's defenses, but Lee held the inner position and could shift his units much faster than Grant. Hotchkiss received a new request from Lee and began preparing detailed maps showing routes to Richmond through the counties of Spotsylvania and Hanover. While Oltmanns and Robinson copied the maps for distribution, Hotchkiss spent most of May 16 sketching the Spotsylvania battlefield in anticipation of requests from Ewell and Lee for after-action maps to accompany their reports.

On May 18 Grant launched one more assault, throwing four Federal corps and heavy artillery at the Confederate left. When the assault failed, Ewell's Second Corps, which had dwindled to 6,000 effectives, rose from the trenches and probed the Federal right flank to determine whether Grant had withdrawn. Ewell crossed the Ny River east of Spotsylvania around 3:30 p.m. and stumbled into Brigadier General Robert O. Tyler's heavy artillery and General Birney's Third Division from Hancock's Second Corps. Ewell's thin force almost became trapped by a Federal counterattack, and he suffered another 900 casualties extricating his corps. Ewell's reconnaissance in force confirmed Lee's suspicions that Grant was preparing to steal another march.

There is no accurate breakdown of casualties during the two-week engagement at Spotsylvania Court House. Confederate combat strength never exceeded 60,000 effectives, with casualties ranging between 9,000 and 10,000. Federal combat strength reached about 111,000, with casualties approaching 18,000. Grant could replace his losses; Lee could not.

The action at Spotsylvania's Bloody Angle was fierce and aptly named. Hotchkiss was on the field moving reinforcements around General Early and witnessed much of the action.

GRANT MOVES SOUTH

During the battle of Spotsylvania, Hotchkiss divided his time between making maps, directing his staff, and visiting the battlefield every day to sketch the fighting and carry messages to the units involved. On the morning of May 21, the battlefield grew quiet, and he woke to find the army preparing to march. "About noon word came that the enemy was moving towards Hanover Junction, so we started down the Telegraph Road as hard as we could," Hotchkiss noted. "We marched until 11 p.m." Once again, by following interior routes organized by Hotchkiss, the Army of Northern Virginia moved faster than Grant, crossed Caroline County, marched over the Telegraph Bridge, and on May 22 bivouacked on the south bank of the North Anna River in a defensive position at Hanover Junction. During the movement, A. P. Hill returned to resume command of the Third Corps, and Early returned to the Second Corps to resume command of his division. A few days later, Ewell admitted that a fall he had suffered at the Bloody Angle had incapacitated him, and he turned command of the Second Corps over to Early.

THE BATTLE OF NORTH ANNA RIVER: MAY 23–27, 1864

By May 23, the Army of Northern Virginia had lightly fortified a line of trenches across Grant's route to Richmond at Hanover Junction. Early, Anderson, and Hill were all in place with their corps when Hancock's Second Corps and Wright's Sixth Corps arrived on the far side of the North Anna River. At Jericho Mills, a few miles upstream from Lee's entrenchments, blueclads from Wright's corps crossed the river, drove off Confederate pickets, and established a bridgehead. Pioneers threw pontoons across the river, and by late afternoon Wright's corps crossed, followed by Warren's Fifth Corps, and formed for an assault against Hill's Third Corps. Hill did not wait to be attacked and sent Major General Cadmus M. Wilcox's division against the enemy. The Confederate assault at Jericho Mills failed, and Wilcox returned to Hill's defensive works with 642 casualties. Having recently obtained 8,000 reinforcements, Lee missed an opportunity because of illness. Grant's faulty dispositions isolated two Federal corps on the south side of the North Anna, and had Lee been on the field the Confederates might have crushed both Wright's and Warren's corps.

Lee did hold a meeting during the afternoon with Hotchkiss and the three corps commanders. Hotchkiss proposed a defensive line along the Virginia Central Railroad and spent the night with Brigadier General Francis H. Smith of the Engineer Corps laying out the defenses. Grant became frustrated and withdrew during the night of May 27. He shifted the army eastward toward Hanovertown and crossed the Pamunkey River. Believing he had flanked Lee, Grant moved the army to Totopotomoy Creek only to find Confederates blocking his crossing. Although no major engagement occurred, Grant was becoming annoyed at Lee's unwillingness to fight in the open and at Meade's inability to break through the Confederate line. After North Anna, Grant began to direct the Army of the Potomac himself.

When Grant's Federals began crossing the North Anna River, Confederate sharpshooters, after harassing the pontoon crews, continued to snipe at the infantry.

Battle of Hanover Junction [Sketchbook Map #8-6]
Hotchkiss titled this phase of the battle of North Anna River the battle of Hanover Junction, and properly so because the major action of the Second Corps occurred on May 27 between the river and the town. He recommended that the Second Corps defend the area around the cut of the Virginia Central Railroad, which contributed to Grant's frustration over the progress of the campaign. Union forces withdrew that night.

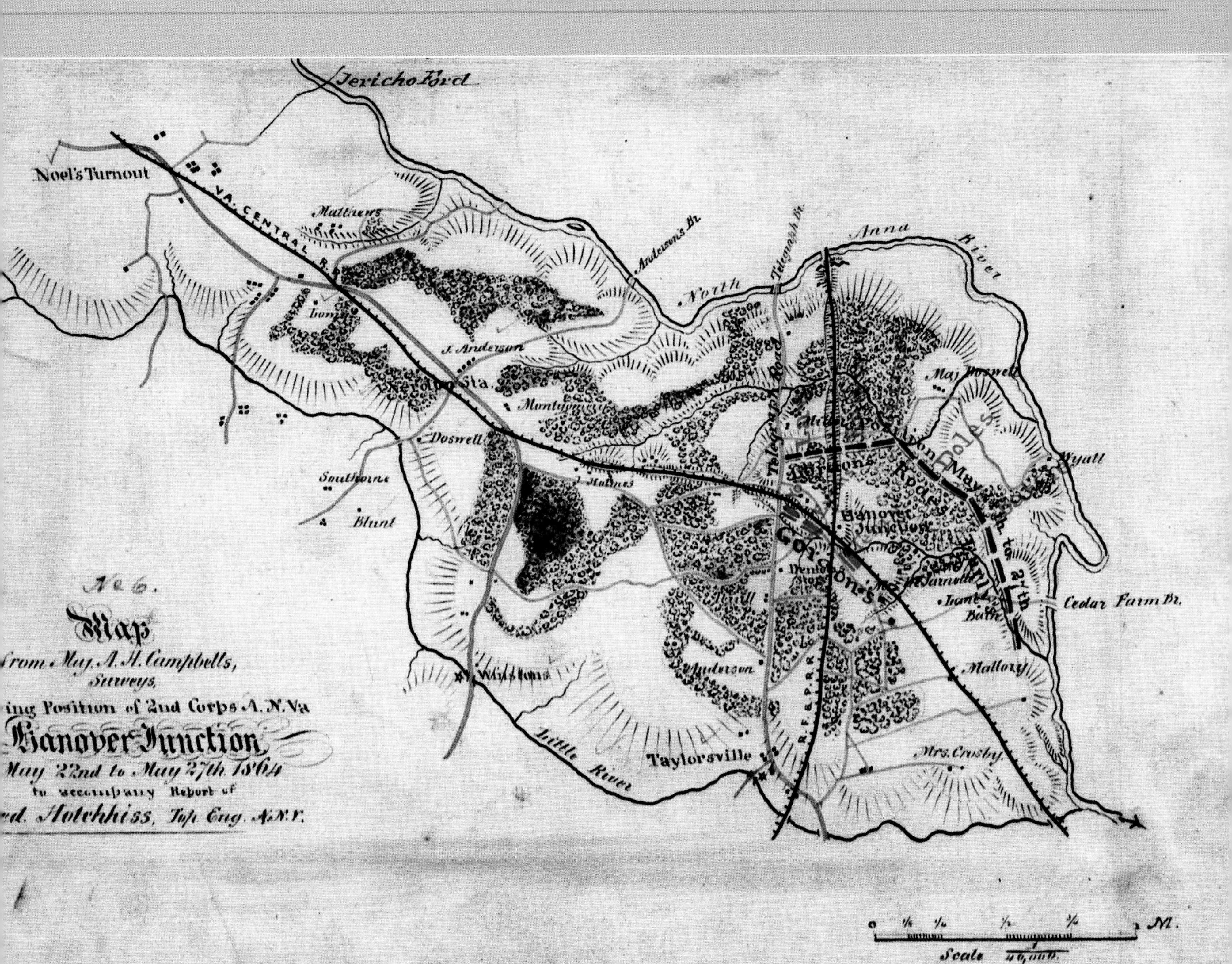

Jericho Ford
Noel's Turnout
VA. CENTRAL R.R.
Matthews
Anderson's Br.
Anna River
North
Telegraph Br.
J. Anderson
Sta.
Maj. Doswell
Doswell
Southorne
Blunt
J. Holmes
Telegraph Road
Hanover Junction
Wyatt
Cedar Farm Br.
No. 6.
Map
from Maj. A. H. Campbells, Surveys,
ing Position of 2nd Corps A. N. Va
Hanover Junction
May 22nd to May 27th 1864
to accompany Report of
ed. Hotchhiss, Top. Eng. A. N. V.
Anderson
R. F. & P. R. R.
Mallory
Little River
Taylorsville
Mrs. Crosby
Scale 1/40,000

TOTOPOTOMOY CREEK: MAY 26–30, 1864

After reaching the Pamunkey River on May 26, Grant's army dwelled in the area for five days, probing for weaknesses in Lee's line. Cavalry clashes occurred at Hanovertown, Enon Church, Ashland, and Hanover Court House. By May 30, the entire Army of the Potomac had taken positions along Totopotomoy Creek. Meade made slight progress in mini assaults against the Confederate line, which Lee deftly parried.

Lee completely understood Grant's game and knew the general would continue to slide south. He fully expected Grant to confront the Confederate line with part of the army while the other part boldly moved on Richmond. On May 27 he detailed Hotchkiss to provide maps from the Totopotomoy to the Chickahominy. Hotchkiss spent May 28 studying the area and correcting obsolete maps kept on file at Lee's headquarters. On May 30 he finished minor revisions to the map of southern Hanover County and another for the vicinity directly east of Richmond. A day later, Grant stopped maneuvering around the Totopotomoy and moved toward Cold Harbor. Lee anticipated the movement and began compressing his lines in the same area.

THE BATTLE OF COLD HARBOR: JUNE 1–3, 1864

While the infantry of both armies marched toward Cold Harbor, General Sheridan's Federal cavalry skirmished with Fitz Lee's Confederate cavalry along the vital road junction at Old Cold Harbor. Sheridan chased off Fitz Lee but was driven away by Confederate infantry. Grant ordered Sheridan back to Old Cold Harbor, and the Federals returned and occupied the area throughout the night. On June 1 Grant's corps began filing into position above the crossroad at Old Cold Harbor, while about a mile to the west Lee's forces began establishing a defensive line around New Cold Harbor.

On hearing that Grant was about to be reinforced by Major General William F. Smith's Eighteenth Corps from General Benjamin F. Butler's stalled Army of the James at Bermuda Hundred, Lee decided his best option would be to attack Sheridan's cavalry on the Federal left before more of Grant's forces arrived to assault the Confederate right. He moved Anderson's First Corps into position opposite Old Cold Harbor and ordered him to attack in the early morning on June 1. Anderson's infantry made slow progress against Sheridan's Spencer rifles and eventually fell back to New Cold Harbor after Wright's Sixth Corps and Smith's Eighteenth Corps arrived on the field.

Through a mix-up of orders, Smith moved his corps into the wrong position and failed to make the necessary adjustment until 4:00 p.m. An hour later, Grant attempted to turn the Confederate right flank by sending six divisions from Wright's corps and Smith's corps against Anderson's corps and Brigadier General Robert F. Hoke's division. The Federals breached part of the Confederate line but were driven back with heavy losses by a furious counterattack.

Once again frustrated by another missed opportunity, Grant ordered Hancock to make an all-night march, take position on Wright's left flank, and attack the Confederate right flank at dawn. While Hancock moved his Second Corps from the Totopotomoy during the night, Lee divined Grant's plan and began strengthening Anderson's right flank with elements from A. P. Hill's Third Corps. Across ten miles of scuffed dirt roads smothered in tropical heat, clouds of choking dust combined with nausea and fatigue to severely slow the pace of Hancock's infantry. Leading elements wearily stumbled into Old Cold Harbor at 6:30 a.m. on June 2 and collapsed. Grant lost his best opportunity to turn Lee's flank and postponed the attack until 5:00 p.m. That afternoon, A. P. Hill's entire corps filed into position opposite Hancock, and Grant postponed the attack for another day.

By evening, the Army of Northern Virginia stretched seven miles from Hundley's Corner to Grapevine Bridge over the Chickahominy River. The Union line ran along a parallel front less than a mile to the west. Lee assumed a defensive posture and capitalized on all the natural defensive features of the landscape. With each corps sharing interconnected flanks with broad fields of frontal and enfilade fire, Lee could not have created a more seamless and flawless defensive alignment.

Having exhausted his patience trying to lure Lee into an open battle, Grant showed his frustration by ordering an attack on June 3. After learning of Grant's plans, many blueclads expected to be killed. In the morning, they pinned scraps of paper bearing their names to their shirts so that their bodies could be identified after they fell. At 4:30 a.m., 40,000 cheering Federals surged across open fields toward concealed enemy positions. The main assault came from Hancock's Second Corps, Wright's Sixth Corps, and Smith's Eighteenth Corps. Suddenly, thousands of rifles raised from behind banked earth and opened in a thunderous crash that no survivor could ever forget. As the Federal front ranks crumbled, a second line followed, giving the Confederates ample time to reload and respond with more deadly volleys. At times, the Federals came within thirty yards of the Confederate works before the firing forced the blueclads to lie down and seek protection in a defile or behind the body of a fallen comrade. The Confederates inflicted enormous devastation on the blueclads, and after thirty minutes of fighting, the Federal assault crumbled except in scattered patches.

Lee's Line of Entrenchments [Map #170]

By June 10, 1864, the Army of Northern Virginia's entrenchments ran from Totopotomoy Creek to the Chickahominy River. When General Grant failed to dislodge the Confederates along the Totopotomoy, he tried at Cold Harbor and failed again. After Grant crossed the Chickahominy and took the Army of the Potomac toward Petersburg, those roughly ten miles of entrenchments were never used again. Hotchkiss did not make a map on the battle of Cold Harbor because on June 13 Lee detached General Early's Second Corps for operations in the Shenandoah Valley.

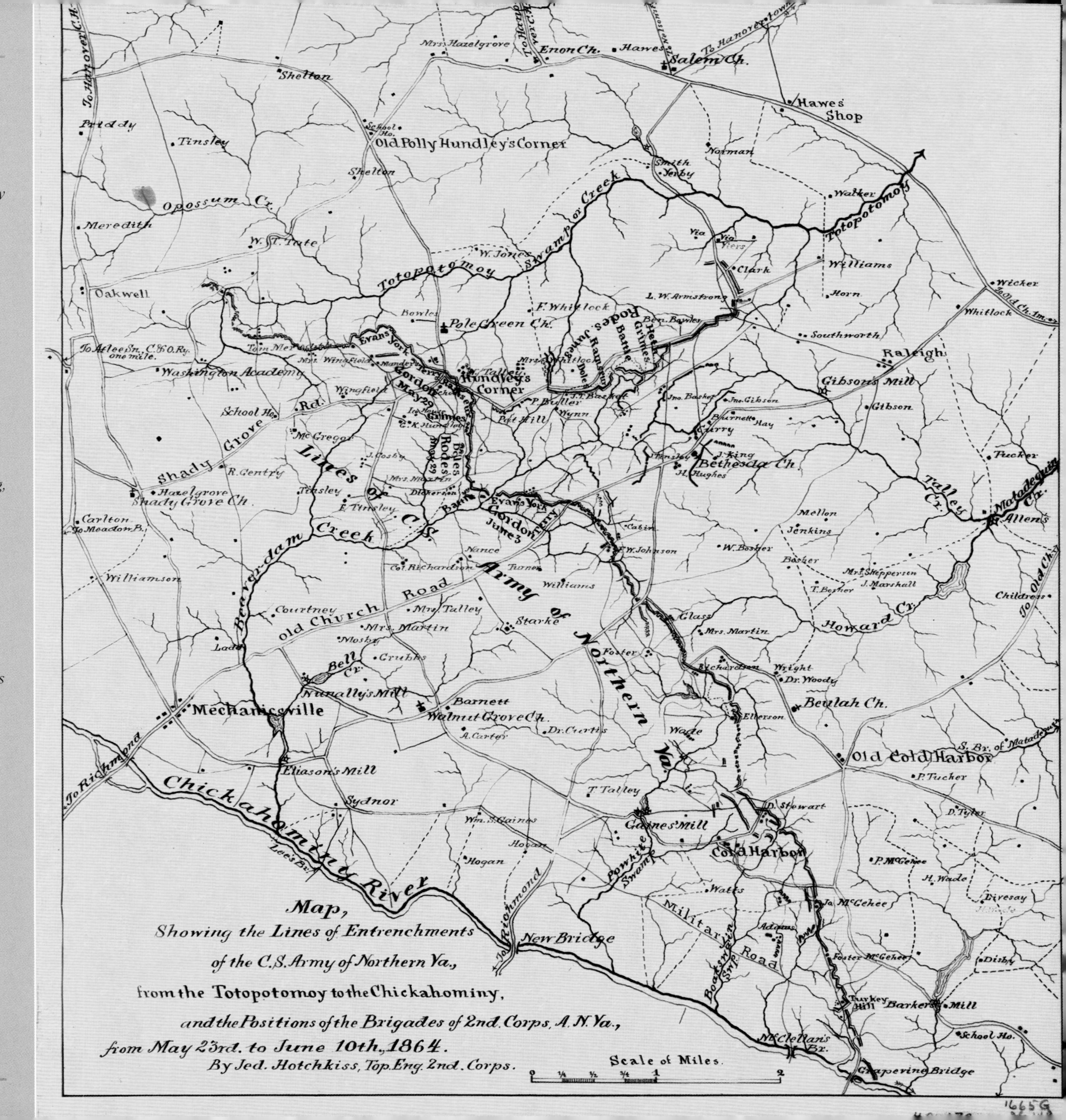

Hancock's men succeeded in making a breach, captured three guns, and turned them on the grayclads before a brigade from Hill's corps counterattacked and regained the position. Brigadier General John Gibbon, commanding Hancock's Second Division, penetrated a lightly defended area of swampy ground and pierced another breach, only to be driven back with huge losses when counterattacked.

General Wright assaulted Anderson's line with three divisions, keeping one in reserve. Although the blueclads temporarily occupied works on the Confederate right, Anderson repulsed fourteen successive assaults.

Smith's Eighteenth Corps attacked across an open field. Thousands of blueclads were cut down by frontal fire from Early's Second Corps and enfilade fire from Anderson's corps. Smith's units became mixed up, and brigade commanders informed Meade of the hopelessness of continuing the attack. Meade ordered Smith to ignore formations and to press forward individually without support. Meade finally relented and the blueclads fell back.

> *"I had seen the dreadful carnage in front of Marye's Hill at Fredericksburg, and on the 'old railroad cut' which Jackson's men held at the Second Manassas; but I had seen nothing to exceed this. It was not war; it was murder."*
>
> —BRIGADIER GENERAL EVANDER McIVOR LAW, C.S.A.

With his Ninth Corps occupying the Federal right opposite the Confederate left, General Burnside pushed two divisions forward to turn Early's flank. The Federals captured some Confederate outposts and stopped. Burnside added another division to the attack, but Meade interceded and ordered a withdrawal to Federal lines. Firing continued sporadically throughout the day and did not cease until nightfall.

Four days passed before Grant requested a truce on June 7 to gather his wounded from the battlefield and bury the dead. By then, many of the wounded that might have been saved had died. Blueclads and grayclads alike, manning trenches in some parts of the line separated by only a hundred yards, began to notice an overwhelming stench emitting from the bodies on the battlefield. Why Grant waited so long to recover his fallen soldiers has never been explained. Years later, when writing his memoirs, Grant said, "I have always regretted that the last assault at Cold Harbor was ever made." For Lee, it would be his last great battle as a field commander.

THE BUTCHER'S BILL

Exact casualty counts during the period from May 28 to June 3 were never made. Approximately 117,000 Federals and 60,000 Confederates participated in the running series of battles immediately before and during Cold Harbor. It is estimated that 13,000 blueclads and about 5,000 grayclads became casualties. More than half of the Union losses occurred on June 3, compared to about 1,200 Confederates.

Although Grant had not decisively won a battle after a month of constant fighting, he was starting to win the battle of attrition. Since the beginning of the Wilderness campaign on May 5, Grant had lost about 50,000 men, or 41 percent of his original strength, to Lee's 32,000 casualties, or 46 percent of his original strength. Grant replaced his men almost as rapidly as they fell, and Lee could not. How much longer it would take Grant to whittle away the rest of Lee's forces simply became a matter of time and resources.

A TEN-DAY INTERREGNUM

During the three-day battle at Cold Harbor, Hotchkiss, Oltmanns, and Robinson worked frantically on maps. Lee expected Grant to attempt a move on Richmond, and every general in the Army of Northern Virginia began asking for maps. "Rode out along the line in the morning," Hotchkiss wrote on June 2, "and took two maps to General Early." The following day, Hotchkiss noted, "I finished a map for Gen. Rodes and took it to him." Afterward, he remained near the battlefield with the Second Corps, sketching Early's defensive line and the surrounding area. Skirmishing continued on June 4 and artillery erupted on occasion, but fighting substantially subsided. Hotchkiss mentioned a small celebration on June 5 when Early received a promotion to lieutenant general.

"Gen. Lee sent for me to make a sketch of our line of battle," Hotchkiss noted on June 7. "I got as far as New Cold Harbor, had some warm work, owing to the constant skirmishing."

Puzzled by Grant's behavior, Lee sent Colonel Smith, his staff engineer, to Hotchkiss with instructions to work on sketches for improving Richmond's fortifications. "It was very unpleasant working through the trenches," Hotchkiss noted. He spent the next two days with Lee's engineers sketching locations for artillery emplacements. The task ended suddenly on June 13 when scouts discovered the Army of the Potomac crossing over the Chickahominy on the Long Bridge. Grant had finally stolen a march on Lee, who put the Army of Northern Virginia in motion in an effort to stop the Federals from flanking Richmond. Lee poured his forces into the trenches at the Confederate capital, only to learn that Grant was still moving south to make a juncture with General Butler's forces at Bermuda Hundred. Lee followed on parallel lines, moving the army into earthworks that had been dug at Petersburg. It became clear to Lee that Grant intended to capture the South Side Railroad, which supplied Petersburg, Richmond, and the Army of Northern Virginia.

Grant's movements became even more troubling when, in mid-June, Lee learned that a strong Federal column had moved up the Shenandoah Valley to threaten Lynchburg, another connecting center for railroads through which supplies flowed to his army and to the Confederate capital on the tracks of the Virginia and Tennessee. Lee had been too engaged with Grant's movements to anticipate the problems developing in the valley. He became furious after learning that Major General David Hunter, commanding Federal forces in the valley, had been on a rampage of devastation, terrorizing the people by burning barns, mills, crops, and the Virginia Military Institute in Lexington.

Although Lee needed every man in the Army of Northern Virginia for the defense of Richmond, he also needed Lynchburg and its railroads. Again, he ignored the conventional maxims of warfare and divided his force in the face of a much stronger enemy, detaching Early's Second Corps for the purpose of driving Hunter's Federals out of the valley.

On June 13 Hotchkiss noted in his journal, "We started at 3 a.m. today. Our whole Corps . . . for some distant expedition."

Early's Second Corps marched twenty-five miles that day. "Corrected map of Louisa County as we went along," wrote Hotchkiss, who never missed an opportunity to advance knowledge of an area for his maps. When the army reached Charlottesville, Hotchkiss knew he was returning to his beloved valley, but he still did not know why.

After two months of constant fighting, Lee's soldiers needed a rest and in the heat of Virginia sought refuge under the trees. Hotchkiss also needed rest, and spent time during the lull to start maps of Louisa County, through which Early's corps would soon be traveling.

CHAPTER SEVEN: JUBAL EARLY'S RAID

"The General had us up by 2:00 a.m. to start to Lynchburg," Hotchkiss wrote on June 17, 1864. Spared from another long march on dry, dusty roads, Early's Second Corps entrained on the Orange and Alexandria Railroad at Charlottesville and reached Lynchburg at 1:00 p.m. Hotchkiss had heard rumors about General Hunter's depredations in the valley, and he understood that to reach Lynchburg the Federals would have passed through Staunton. He naturally worried about his family, home, and property. Several days later, he received a letter from his wife reporting that all was safe and well.

Early's return to the valley came as a great relief to the inhabitants. When Hunter began his campaign of destruction in May, the only Confederate force in the valley was Brigadier General John D. Imboden's 1,600-man cavalry brigade. Major General John C. Breckinridge, recently placed in charge of the Department of Western Virginia, commanded a 6,500-man infantry division, but the troops were 155 miles away. Before Lee decided to dispatch Early's corps to Lynchburg, he sent a request to the war department asking that Imboden and Breckinridge be attached to the Army of Northern Virginia so that their forces could be unified in a common command. After Lee learned that two Federal armies had been organized under Hunter, he detached Early's corps, which on arrival at Lynchburg assimilated Breckinridge's and Imboden's commands.

Early retained the same staff that had served with Jackson and Ewell. Hotchkiss's status had never changed. He still drew a captain's pay and remained a civilian employee. Shortly after arriving at Lynchburg, Early moved Hotchkiss and his topographers into an office in the city and put them to work making maps of the area. Later in the day, Hotchkiss distributed copies to generals Rodes, Gordon, Breckinridge, and Ramseur. The topographical crew spent about a day in the office before Early issued orders on June 18 for the corps to move at daybreak.

THE BATTLE OF LYNCHBURG: JUNE 17–18, 1864

When Lee sent the Second Corps to Lynchburg, he instructed Early to punish Hunter, chase the Federals out of the valley, and with swift marches cross the Potomac and threaten Washington, D.C. On June 16 Early telegraphed from Charlottesville imploring General Breckinridge at Lynchburg to expend every effort possible to block Hunter's force from entering Lynchburg for one more day. Hunter's command consisted of 18,000 blueclads with thirty guns, which could have rolled right through Breckinridge's 2,000-man brigade with little effort.

Late on June 17, Hunter's corps appeared about half a mile north of Lynchburg and began shelling Breckinridge's thinly distributed defensive line rimming the outer edge of town. Early's vanguard arrived during the afternoon, occupied defensive positions, and began responding with a few pieces of artillery.

On the morning of June 18, Hunter advanced cautiously, at first believing that Lynchburg would be easy to capture. He paused to reconsider when he observed the outskirts of the town bristling with 8,000 Confederate rifles. He delivered a few probing attacks, all of which the grayclads readily repulsed. Hunter hesitated throughout the morning and finally decided upon a flanking movement.

Around noon, Early became impatient and attacked. Completely unprepared for such audacity, and with his flanking movement still stalled, Hunter recalled his men with "great promptitude" and, according to Colonel David H. Strother, Hunter's chief of staff, "immediately faced them about and, waving his sword, led them back to their original position." After flagrantly underestimating the size of Early's force, Hunter abruptly reversed his original opinion and grossly overestimated the size of the Confederate force. After losing about 1,000 men, he also lost heart. Being thoroughly demoralized, Hunter fled into western Virginia and never stopped until he reached Parkersburg on the Ohio River. For more than a week, the Union war department did not hear from Hunter and feared the Confederates had somehow swallowed up his entire force.

EARLY MOVES NORTH

"The enemy retreated last night," wrote Hotchkiss on June 19, "and this morning we followed them, Ramseur in advance, along the Salem Road." With Early on his heels, Hunter never moved so fast. Unacquainted with such strenuous marching, hundreds of exhausted blueclads collapsed on the road and were captured. After Hunter's corps disappeared into West Virginia, Early started north, marching down the valley toward Maryland. With every passing mile, Early's soldiers became more shocked by the devastation wrought by Hunter.

Major General Robert Ransom Jr. (1828–92) *served under General Breckinridge and later under General Early during the 1864 Shenandoah Valley campaign. He fell ill in August after the attack on Washington, D.C., and never returned to the valley.*

Battle of Lynchburg [Sketchbook Map #8-9A]
There might never have been a map of the battle of Lynchburg had Hotchkiss not arrived in time to collect the information required to make one. The map shows Breckinridge's position on June 17, when he attempted to defend the outskirts of Lynchburg with barely 2,000 men. After Early arrived, Gordon, Ramseur, and Breckinridge, with the aid of Brigadier General John McCausland's cavalry, advanced and formed a new line that ended the Federal threat at Lynchburg.

JUBAL ANDERSON EARLY (1816–94)

Born in Franklin County, Virginia, Early graduated from West Point in 1837, ranking eighteenth in a class of fifty, and joined the artillery. He resigned in 1838 to become a lawyer and a Whig in Virginia's house of delegates. During the Mexican War, he served as a major of Virginia volunteers. In April 1861 he voted against secession in the Virginia convention, but afterward promptly joined the Confederate army as a colonel of the Twenty-fourth Virginia Infantry. He led the regiment at First Manassas, after which the war department raised "Old Jube," as his men called him, to brigadier general.

Still a bachelor at the outbreak of the war, Early stood about six feet tall, weighed about 170 pounds, and suffered from arthritis, which did not improve his disposition. He wore a long black beard and had sharp black eyes that Hotchkiss said "looked right through you." The mapmaker did not approve of Early's irreligious nature but was amused by his mordant wit and admired the general's underlying warmth and loyalty, which he usually concealed from others. In his *Reminiscences of the Civil War*, John B. Gordon recalled Early's lack of decorum and referred to the general's bad temper and habitual swearing as "a pungent style of commenting on things he did not like."

Early rapidly established himself as a determined fighter. Wounded during the Peninsula campaign, he returned in time to join Stonewall Jackson at Malvern Hill and lead a brigade at Cedar Mountain and Second Manassas. At Antietam, after General Lawton fell, Early assumed command of what had previously been General Ewell's division in Jackson's corps. Old Jube fought in all the major battles of the Army of Northern Virginia and continued to lead his division as a major general at Chancellorsville, Gettysburg, and the Wilderness. On May 29, 1864, he succeeded Ewell and led the Second Corps at Cold Harbor. In June, Lee detached the Second Corps, elevated Early to lieutenant general, and sent him into the Shenandoah Valley with instructions to raid Washington, D.C. This marked the beginning of Early's famous Washington raid in the months before the collapse of the Confederacy.

According to Douglas Southall Freeman, Early gradually developed from an impetuous commander into one "second only to Jackson himself," which became manifest during the Washington raid and through the final battles in the valley. After the war, Early made his way to Mexico in disguise and then to Canada. He eventually returned to Lynchburg, resumed his law practice, and became president of the Southern Historical Society. Always loyal to Virginia and the South, Early died in Lynchburg on March 2, 1894, "unreconstructed" to the end.

"Houses had been burned, and helpless women and children left without shelter," Early wrote in his memoirs. "The country had been stripped of provisions and many families left without a morsel to eat. Furniture and bedding had been cut to pieces, and old men and women robbed of all the clothing they had except that on their backs." Such sights made Early and his grayclads unusually vindictive.

Hunter's hasty departure opened the entire corridor of the valley to Early. At daylight on June 23, 1864, 10,000 Confederate infantry in four divisions, half of them barefoot, with 4,000 cavalry in four brigades and forty guns, filed onto the roads leading to Lexington. Breckinridge, Gordon, Rodes, and Ramseur, all tested veterans, each commanded a division of infantry. Major General Robert Ransom Jr. commanded the cavalry division. Early did not delude himself into believing that his mini campaign would lead to a victorious conclusion to the war. He understood his task, which was to alarm the North, thus compelling Grant to shed some of his strength, thereby alleviating pressure on Lee's army defending the Richmond–Petersburg entrenchments.

Early wasted no time marching down the valley and on June 27 reached Staunton and formed his four infantry divisions into two corps of two divisions each for the invasion of the North. Rodes commanded one corps and turned his division over to thirty-five-year-old Brigadier General Cullen A. Battle. Breckinridge commanded the other corps and turned his division over to forty-one-year-old Brigadier General John Echols.

"We rode into Winchester," Hotchkiss wrote on July 2. "The people . . . were so glad to see us and as loyal as ever." While Hotchkiss rested and tinkered with his maps, Breckinridge started for Martinsburg with Gordon and on July 3 chased away the Federal garrison while Rodes and Ramseur marched through Charlestown and drove the enemy

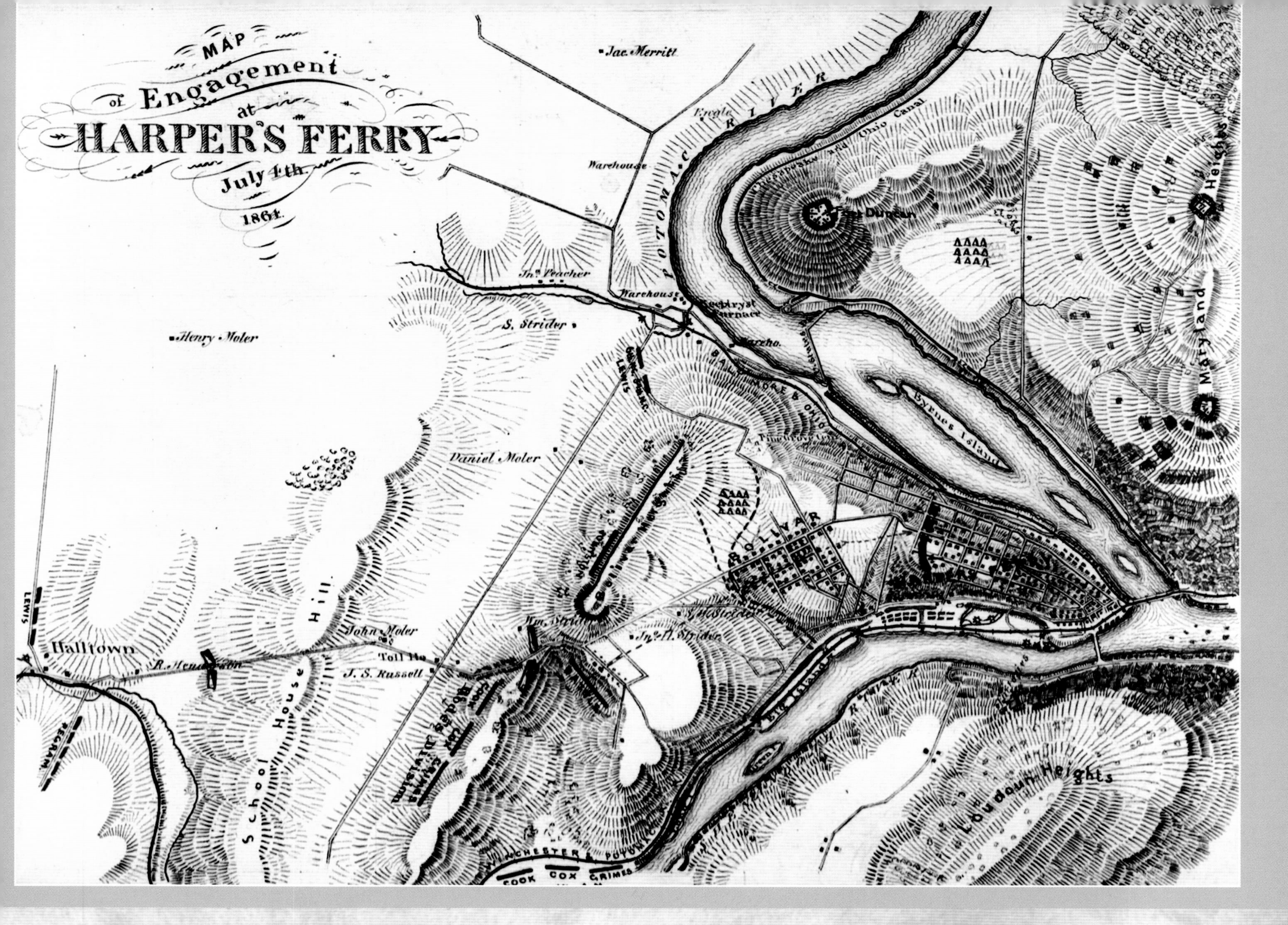

into Harpers Ferry. Hotchkiss went with Rodes and observed the skirmish from Bolivar Heights, which overlooked Harpers Ferry. Major General Franz Sigel evacuated the town on the night of July 4 and moved his command across the river to Maryland Heights. Rodes collected tons of supplies left behind by Sigel and departed from the area on the following day.

Engagement at Harpers Ferry [Sketchbook Map #8-12A]

From his perch on Bolivar Heights, Hotchkiss enjoyed a beautiful view of the surrounding mountains, the line of Federal trenches around the perimeter of Bolivar, and the town of Harpers Ferry at the confluence of the Shenandoah and Potomac rivers. After Sigel's Federals escaped across the bridge to Maryland Heights, Early decided to not waste time fighting and ordered Rodes to take his divisions to Shepherdstown and cross the Potomac.

OPERATIONS IN MARYLAND

"Today we crossed the Potomac and went to Sharpsburg," Hotchkiss wrote on July 6. He then spent the next few hours making maps for Major Harman so the chief quartermaster would not get lost. The time spent on the old Antietam battlefield made it possible for Early's four infantry divisions to assemble and rest. The Confederates waited for the return of General McCausland's cavalry, which had stopped at Hagerstown and levied a requisition of $20,000. Two days later, the army marched through Boonsboro and on July 9 camped outside Frederick. Early went into the city the following day and levied another requisition—this time for $200,000.

Not until July 5 did either Halleck or Grant believe that Early's small army intended to cross the Potomac and invade Maryland. The next day, Grant detached Major General James B. Ricketts's 5,000-man division from Wright's Sixth Corps and sent them to Baltimore accompanied by 3,000 dismounted cavalry. The cavalry went on to Washington, and Ricketts joined Major General Lewis Wallace's Eighth Corps near Frederick. On July 9 Wallace's troops collided with Early's corps along the Monocacy River southeast of Frederick.

Major General Robert Emmett Rodes (1829–64), after being wounded at Seven Pines and again at Antietam, died at the third battle of Winchester on September 19, 1864, during General Early's Shenandoah campaign.

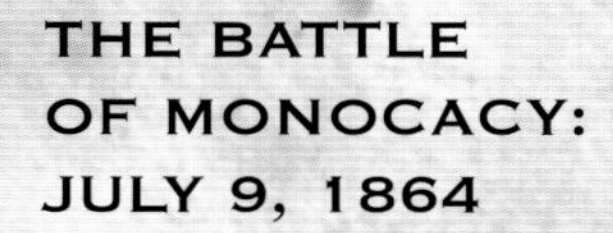

THE BATTLE OF MONOCACY: JULY 9, 1864

On July 8 General Wallace anticipated Early's next move and fell back to a defensive position across the Monocacy River. He chose the site because of two important crossroads, one leading to Baltimore and the other to Washington, which he predicted would be Confederate objectives.

The following morning, Early found the roads blocked across the Monocacy by Wallace's corps and Ricketts's division. While Rodes threatened the Federal right, Ramseur pressed back Federal skirmishers at the Monocacy Junction railroad bridge. Like every Confederate battle near the Mason-Dixon Line, there always seemed to be fields of corn and wheat, and from those on the far side of the river the Federals repulsed three Confederate charges. Early assessed the battlefield and at 4:00 p.m. decided to send Gordon's division against Wallace's left flank while Ramseur assaulted the right flank. Gordon crossed the river and assaulted Ricketts's regiments while Echols's brigades struck Wallace's center. Although Gordon found it difficult to penetrate an obstinate defensive stand by Ricketts's veterans, the fourth Confederate charge broke the Federal line. Wallace ordered a withdrawal and retreated toward Baltimore.

The one-day battle cost Wallace 1,880 men and Early 700. The day lost at the Monocacy also cost Early his one and only chance of attacking Washington before elements from Wright's Sixth Corps and the First Division from General William H. Emory's Nineteenth Corps reached the capital's defensive works. Grant had not wanted to detach Wright, whose divisions began disembarking at midday on July 11 and had moved that evening into the city's outlying fortifications. He had, however, expected General Hunter's force to be in the area and not on the far side of West Virginia. Hunter had lost all contact with Early and showed no interest in finding him.

Battle of Monocacy
[Sketchbook Map #8-15]

Hotchkiss described the invasion of Maryland as "exhausting" because he "marched with the army during the day and worked on maps all night." The victory at Monocacy rejuvenated Hotchkiss. He circulated through the battlefield with renewed enthusiasm on the morning of July 10 to sketch a map of the action. He didn't get another chance to work on it until August 8 because Early's movements never stopped. Had the Confederates been blocked or defeated at Monocacy River, the main purpose of the campaign would have been nullified.

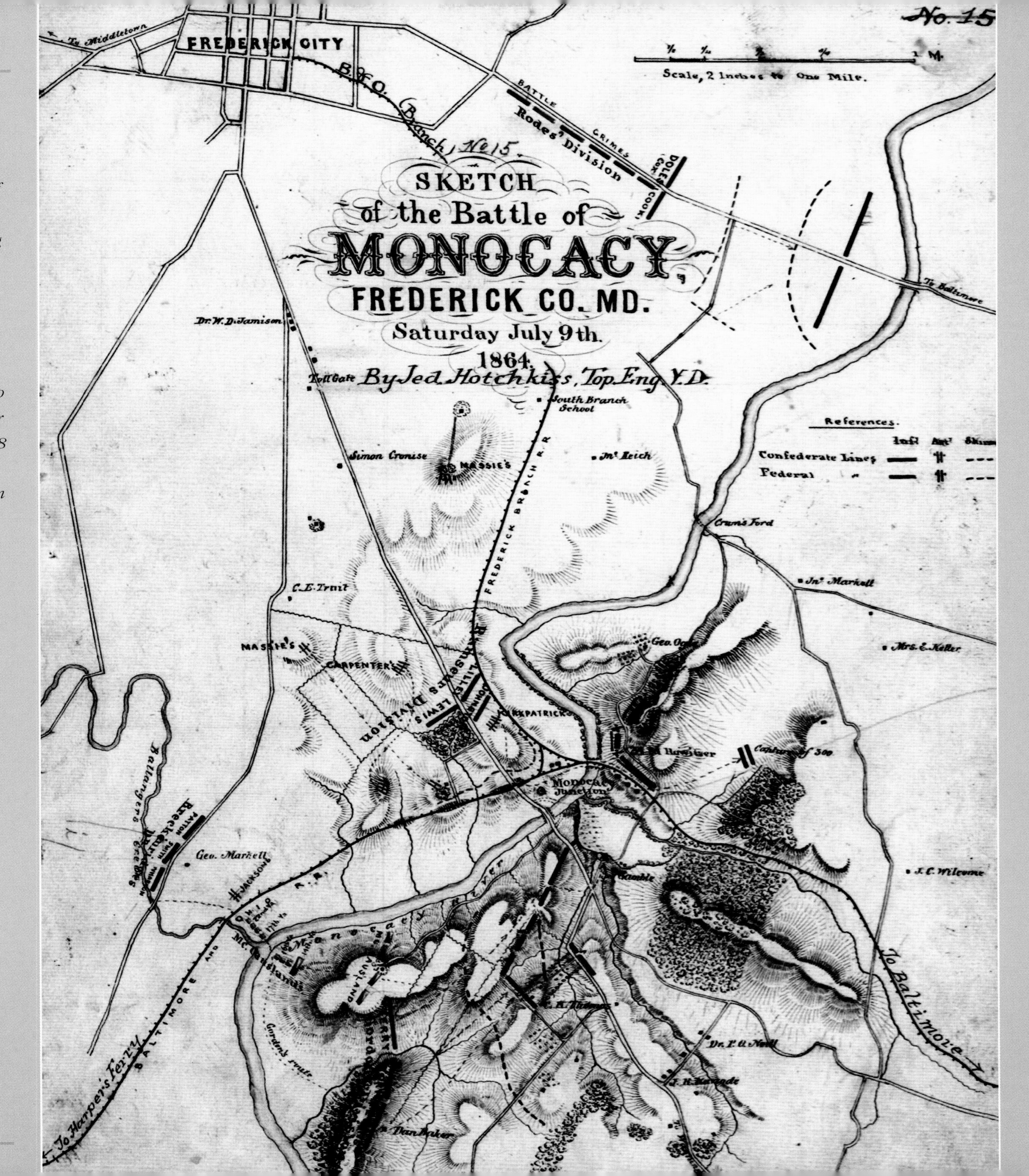

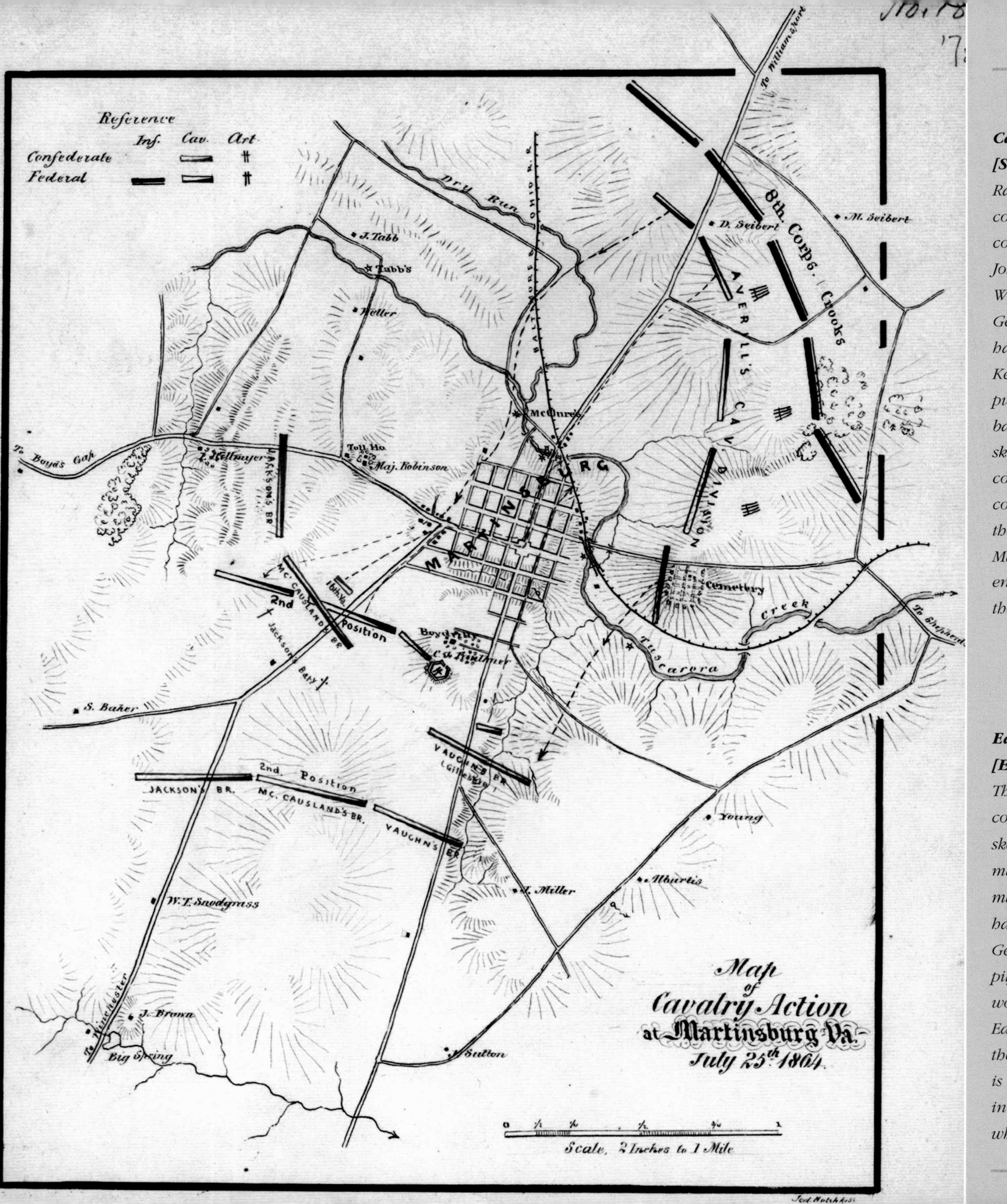

Cavalry Action at Martinsburg [Sketchbook Map #8-17A]

Ransom's cavalry division consisted of three brigades commanded by Brigadier General John McCausland, Colonel William L. Jackson, and Brigadier General John Vaughn. Averell had not led his cavalry well at Kernstown, but succeeded in pushing the Confederate cavalry back at Martinsburg. The skirmish on July 25, however, convinced Crook that Early's corps would not be far behind the cavalry, so he retired into Maryland, thus leaving the entire valley in possession of the Confederates.

Early's Washington Raid [Envelope Map #110]

The map in the Hotchkiss collection came from a rough sketch he or his staff probably made from a commercial city map of Washington, which may have come from Postmaster General Montgomery Blair's pilfered collection. Hotchkiss would have intended to use it for Early's after-action report, which the general never requested. There is also a similar but smaller map in the Hotchkiss collection, both of which contain the same detail.

ON TO WASHINGTON

The Confederates regrouped during the night, and on the morning of July 10 the weary column passed through Urbana on their way to Washington. Early sent a detachment of cavalry to threaten Baltimore and left another detachment on the Monocacy to protect his communications.

About noon on July 11, Confederate troops began filing into Silver Spring, Maryland, about five miles north of Washington and two miles from Fort Stevens, which stood beside the main road leading into the capital. The heat and choking dust from the march had debilitated the grayclads and they collapsed on the ground while Early sent scouts ahead to reconnoiter the capital's defensive works. Early studied the nearest forts with his glass and observed reinforcements filing into the parapets. After spending the balance of the day reconnoitering and resting his troops, Early ordered an assault at daybreak. He postponed the attack during the night after scouts reported the arrival of the remainder of Wright's Sixth Corps. In the morning, Early sent skirmishers toward Fort Stevens to test Federal strength and decided against an assault. "We spent the day in front of Washington," Hotchkiss noted, but he did not mention that the Confederates had spent the day plundering homes and businesses. A few vindictive grayclads looted and burned the mansion of Postmaster General Montgomery Blair, who was on a fishing trip in Pennsylvania. Hotchkiss also entered the mansion and ransacked Blair's map collection.

At nightfall, loaded down with supplies and booty, Early's army headed back up the Potomac by the same route they had come. On July 14, the army crossed the river at White's Ford near Leesburg, and three days later passed through Snicker's Gap and waded across the Shenandoah. During the withdrawal, Hotchkiss put Robinson and Oltmanns to work on preparing maps of Early's operations. Wright led an ineffectual Federal pursuit without manifesting a willingness to engage in battle, despite the fact that the Confederates had encumbered their withdrawal with 1,000 head of cattle. Early had hoped to hold Washington under siege long enough for Brigadier General Bradley T. Johnson's cavalry brigade to free 20,000 Confederates imprisoned at Point Lookout, Maryland, but the effort failed. One grayclad, however, summed up the expedition by declaring, "We haven't taken Washington, but we've scared Abe Lincoln like hell!" In this regard, Early fulfilled his mission by relieving a modest amount of pressure on the Petersburg lines.

THE SECOND BATTLE OF KERNSTOWN: JULY 23–24, 1864

Early returned to the Shenandoah Valley and on July 19 learned that General Wright, with Brigadier General William H. Emory's divisions from the Nineteenth Corps, had established contact with Major General George Crook, who was leading Hunter's corps back to the valley. Early also learned that Brigadier General William W. Averell, commanding the Second Cavalry Division, had a strong mixed brigade that included infantry and had begun advancing on Winchester from Martinsburg. Ramseur attempted to check Averell on July 20 but fell back. Hotchkiss called Ramseur's defeat "a disgraceful affair" because the general "lost four pieces of artillery and 400 men." Averell entered Winchester the following day, and Crook arrived on July 22.

The situation became complicated for the Federals because Averell, Crook, Emory, and Wright all served different military commanders. General Grant believed that Early was returning to General Lee's army and recalled Wright, who on the evening of July 20 began preparations to return his command to Washington. Grant also believed that Hunter could deal with Early, although Hunter had habitually avoided engaging the Confederates for fear of being captured and punished for his depredations in the valley. On July 23 Wright's corps reached Washington and waited while Grant and Halleck deliberated about whether to send Wright back to the valley or return the corps to the Army of the Potomac. That same indecisiveness did not exist among the Confederates. On July 23 Early issued orders to drive Crook out of Kernstown, where the Federals were camped.

Early had retreated to Strasburg, twenty miles south of Winchester, when he decided to assault Crook. The Federal commander had just reorganized his 12,000-man infantry into three divisions. At Kernstown, Crook also had 1,500 troopers in Averell's cavalry division. On the same day, Crook sent Averell toward Strasburg to reconnoiter. Averell collided with Ransom's cavalry and found the pike teeming with Confederates marching toward Winchester. After a brief skirmish, Averell returned to Kernstown and informed Crook.

On the morning of July 24, Early sent Gordon's division to probe Crook's strength. The demonstration brought on heavy fighting. Echols's division from Breckinridge's corps struck Crook's exposed left flank and repulsed the Federal cavalry. Gordon and Echols charged in two perfectly coordinated wings that broke the Union line. During the ensuing rout, Ransom's cavalry attacked Crook's supply train and captured seventy-two wagons laden with provisions and ammunition.

Crook fled north, pursued by Ransom's cavalry. On July 25 Crook made a feeble stand against Ransom at Martinsburg, quickly withdrew after a skirmish that partly occurred on the streets of the town, and retired into Maryland.

CONFEDERATE REPRISALS

On July 25, with Early again on the march into Maryland, Hotchkiss wrote, "I got drawing boards set up; got chairs, etc. Rodes marched to Bunker's Hill, in advance, followed by Ramseur and Breckinridge." Unable to predict how far north Early intended to go, Hotchkiss put Robinson and Oltmanns to work on copying maps and making a reduction of Adams County, Pennsylvania. General McCausland asked for a map of western Maryland and on July 28 put his cavalry brigade, along with Johnson's brigade, on the road to Pennsylvania. Early followed with infantry and tore up the recently repaired tracks of the Baltimore and Ohio at Martinsburg.

Under direct orders from Early and in reprisal of Hunter's destruction of property in the valley, McCausland and Johnson reached the outskirts of Chambersburg, Pennsylvania, at 3:00 a.m. on July 30 with 2,600 cavalry and announced their arrival by firing three artillery rounds. Five hundred Confederates rode into town, where McCausland read Early's instructions and levied an assessment of $100,000 in gold or $500,000 in greenbacks to indemnify three Virginians whose homes had been torched by Hunter's order. McCausland warned that if townsfolk refused to pay the levy in six hours, Chambersburg would be set on fire. At 9:00 a.m., when it appeared that the inhabitants either could not or would not comply with the edict, McCausland ordered the town's 3,000 citizens to evacuate. Confederates began looting stores and businesses while others broke into private homes to plunder silverware, jewelry, cash, and liquor, with which a number of grayclads became obnoxiously drunk.

With the looting underway, firebrands torched the courthouse, town hall, and the town's warehouse. Ten minutes later, fire and smoke smothered Chambersburg's business district. Townsfolk fled to the outskirts carrying a few possessions and watched in horror as their homes caught fire. Several citizens paid cash to not have their homes burned, but two-thirds of the town became engulfed in flames without any distinction between those who paid tribute and those who refused.

The Confederates withdrew at 1:00 p.m., leaving 400 buildings smoldering. The cost of the damage exceeded $1.5 million. Early's retaliation would eventually cost the inhabitants of the valley much more.

Cavalry Operations at Chambersburg [Sketchbook Map #8-18A]
With the Federals driven from the valley in late July, Early's cavalry went on a rampage into Pennsylvania and burned Chambersburg. McCausland and Johnson went into West Virginia and bivouacked at Moorefield. Surprised by Averell's cavalry at dawn on August 7, the Confederates were routed, losing 420 men, 400 horses, and four guns. General Johnson was also captured, but escaped. Hotchkiss created a map to show the routes of the twelve-day raid.

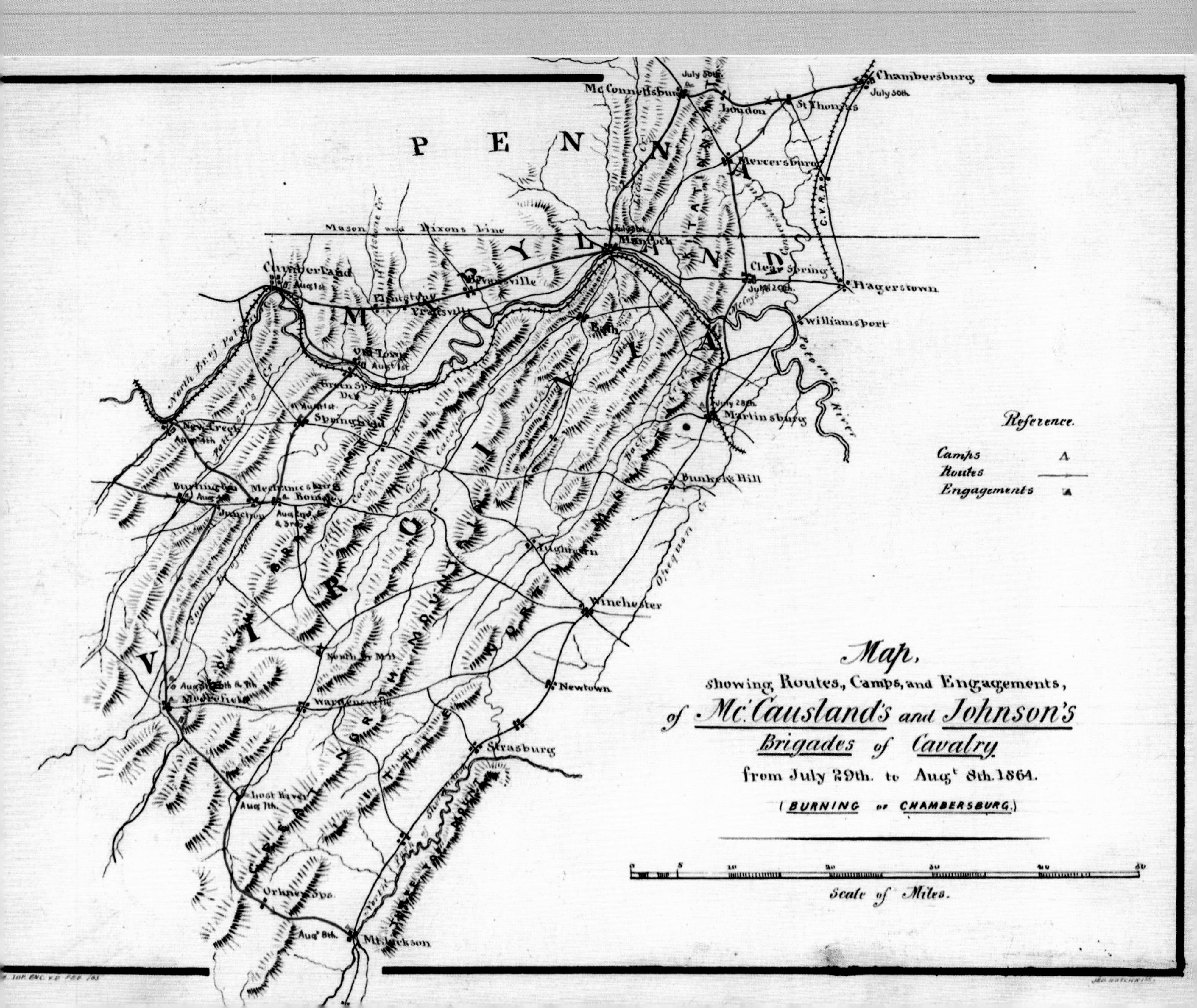
Map,
showing Routes, Camps, and Engagements,
of Mc Causland's and Johnson's
Brigades of Cavalry
from July 29th. to Augt 8th. 1864.
(BURNING of CHAMBERSBURG.)
Reference.
Camps
Routes
Engagements
Scale of Miles.
PENNSYLVANIA
MARYLAND
VIRGINIA
Chambersburg
McConnellsburg
Mercersburg
Mason and Dixons Line
Hancock
Cumberland
Clear Spring
Hagerstown
Williamsport
Martinsburg
Bunker's Hill
Winchester
Newtown
Strasburg
Moorefield
Springfield
Orkney Sps.
Mt. Jackson

JOHN McCAUSLAND (1836–1927)

Born in St. Louis and educated at the Virginia Military Institute, McCausland graduated in 1857 at the top of his class. After graduating from the University of Virginia, he returned to V.M.I. to teach mathematics.

He recruited the Thirty-sixth Virginia Infantry in 1861 and, as the unit's colonel, led the regiment in western Virginia. Transferred to Kentucky and Tennessee, McCausland eventually returned to Virginia as a brigade commander and served under General Echols. Elevated to brigadier general on May 18, 1864, he succeeded to the command of Albert Jenkins's cavalry brigade after the latter's death.

McCausland led cavalry for most of the remainder of the war, campaigning actively and with distinction in General Early's 1864 Shenandoah Valley campaign. He is best remembered for his raid into Pennsylvania and the burning of Chambersburg after the citizens refused to pay reparations for General Hunter's destruction of private property in the valley.

After operations in the Shenandoah Valley subsided, McCausland joined General Lee at Petersburg. He led his cavalry at Five Forks and cut his way through Federal lines to Appomattox before the final surrender. After receiving his parole, McCausland spent two years in Mexico and Europe as an expatriate before returning to the South and acquiring a 6,000-acre tract in West Virginia. When he died at the age of ninety-one in 1927, only one other general of the former Confederacy was still alive.

GRANT FINDS A GENERAL

Frustrated with the situation in the Shenandoah Valley and annoyed by the poor performance of Hunter, General Grant appealed to President Lincoln. "All I ask," he said grimly, "is that one general officer, in whom I and yourself have confidence, should command the whole [Valley District]." Grant had already decided on putting thirty-three-year-old Philip H. "Little Phil" Sheridan, commander of the Army of the Potomac's Cavalry Corps, in charge of the district but knew that opposition would come from Secretary of War Stanton and from his own chief of staff, General Halleck. Grant did not intend to debate the issue. He wanted the four independent departments in the district merged into one and Hunter to be removed from command. "I want Sheridan put in command of all the troops in the field, with instructions to put himself south of the enemy and follow him to the death." Lincoln finally agreed, Grant worked out the details, and on August 7 the war department officially created the Middle Military District with Sheridan in command and, as Grant wished, merged the four military departments.

Born in Albany, New York, in 1831 to Irish immigrant parents, Sheridan falsified his date of birth by a year to gain early entrance to West Point. Suspended a year for losing his temper and chasing a cadet officer with a bayonet, Sheridan graduated a year late, thirty-fourth in a class of fifty-two. Until the outbreak of the rebellion, Sheridan fought hostile Indians in the Northwest. During the first year of the war, he became a captain in the Quartermaster Corps while all his colleagues rose in the ranks and became colonels and generals. When given an opportunity in May 1862 to become colonel of the Second Michigan Cavalry, Little Phil's career swiftly changed. In June he received command of a brigade, and in September he assumed command of the Eleventh

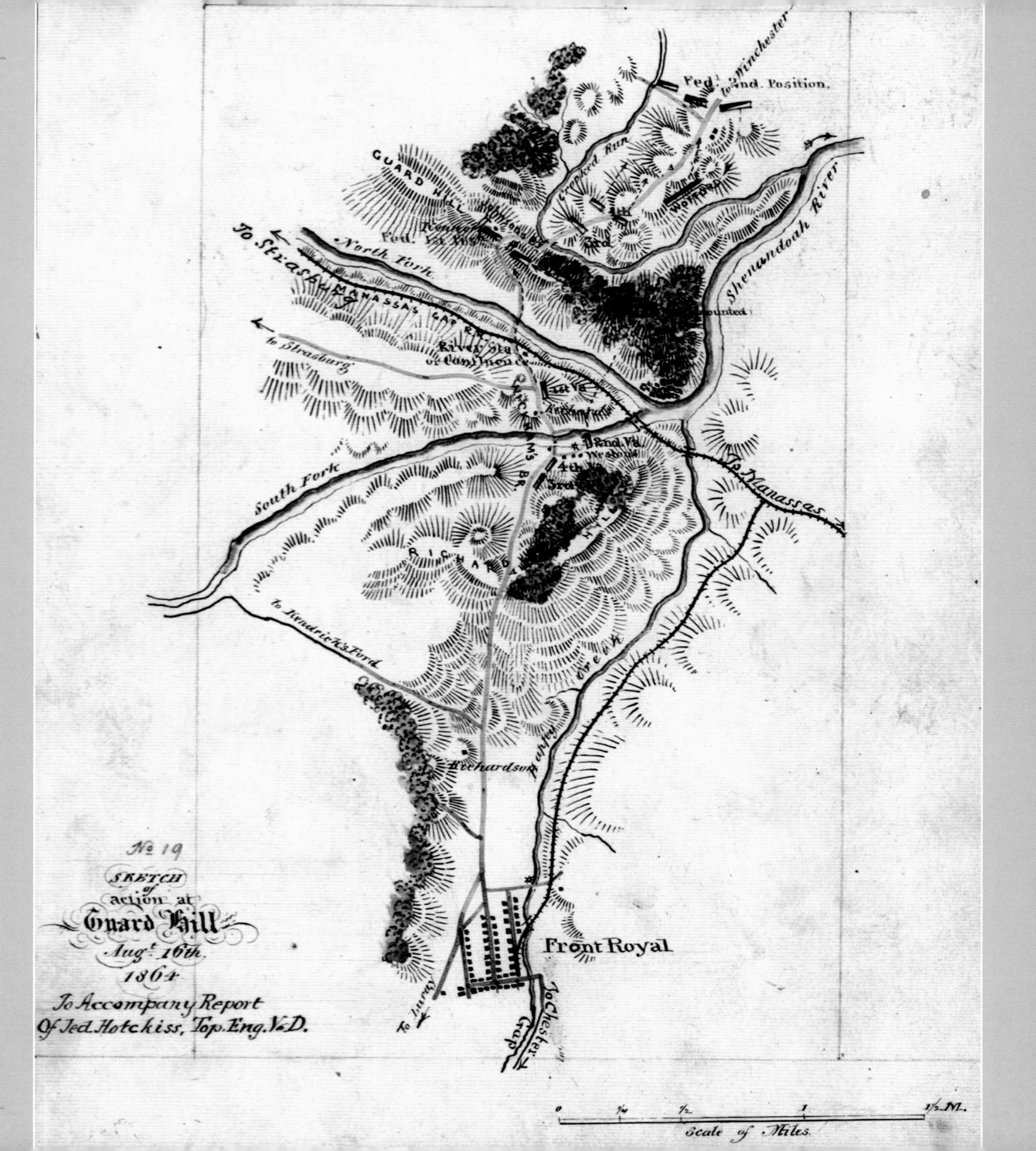

Action at Guard Hill
[Sketchbook Map #8-19]

Hotchkiss summed up the fight at Guard Hill, also referred to as the engagement at Cedarville or Front Royal, by writing, "Anderson had a fight with the Yankee cavalry at Guard Hill, in which he got the worst of it." Hotchkiss would not normally make maps of small engagements, but he now had a staff to keep busy. Hotchkiss had previously recorded every detail of the topography. After Oltmanns copied the area involved from the Shenandoah Valley map, Hotchkiss plotted the positions and movements of the combatants.

Division in the Army of the Ohio. A month later, he fought with distinction at Perryville, Kentucky. In December 1862, after leading the Third Division, Sixteenth Corps at Stones River, Sheridan won his second star. He fell under Grant's wing during the battle of Chattanooga, and when Grant became general-in-chief, he moved Sheridan into the Army of the Potomac and put him in charge of the cavalry corps. Having recognized Sheridan's aggressiveness and natural leadership as both an infantry and cavalry commander, Grant found the right man to regain control of the Shenandoah Valley.

A CALM BEFORE THE STORM

From the end of July and the early weeks of August, Hotchkiss wandered systematically through the lower valley with Robinson and Oltmanns, virtually unimpeded by Federal interference. The topographical staff made maps of Early's Washington raid and many of the skirmishes that occurred during the campaign. He also added more detail to his map of the Shenandoah Valley and created new maps of the lower valley. On August 5 he rode with Early to Sharpsburg because the general wanted a sketch showing the position of his division on the Antietam battlefield. When Federals began to concentrate near Harpers Ferry on August 6, most of Early's forces were north of the Potomac River loading wagons with freshly cut grain. Warned by scouts of the arrival of Federal reinforcements, Early pulled back to the area of Martinsburg and Bunker Hill to avoid being cut off from his communications.

While Early collected supplies and skirmished with Federal cavalry, Sheridan organized the newly formed Army of the Shenandoah into a force of more than 40,000 troops. The new army included Wright's Sixth Corps, Crook's Eighth Corps, two divisions of Emory's Nineteenth Corps, and three cavalry divisions, two of which had served under Sheridan in the Army of the Potomac. Early's pint-sized Army of the Valley consisted of 12,000 men organized into two corps of two divisions each, two artillery battalions, and two undersized cavalry divisions.

About a week after Sheridan arrived at Harpers Ferry, the Federals advanced up the valley. On August 12 Early withdrew to a naturally strong position on Fishers Hill, about two miles south of Strasburg, and skirmished for three days. Hotchkiss spent August 14 "under a tree by the roadside" sketching the action, which became the beginning of Sketchbook Map #8-26 (see page 149).

While skirmishing, Early noticed that the Federal army had greatly increased in size and sent a message to Lee requesting reinforcements. Lee diluted his forces at Petersburg and detached Richard Anderson with an infantry division, a cavalry division, and an artillery battalion. Anderson arrived at Front Royal on the afternoon of August 14 and put part of his command on Guard Hill to defend one of the two main roads to Winchester.

THE BATTLE OF GUARD HILL: AUGUST 16, 1864

The arrival of Anderson's force changed Sheridan's mind about assaulting the Confederates on Fishers Hill. He grew concerned on August 16 over the possibility that his communications could be cut and his rear or flank assaulted by Anderson. Uncertain of the size of Anderson's force, Sheridan ordered Brigadier General Wesley Merritt's cavalry division toward Front Royal to reconnoiter the roads. Anderson parried by sending Brigadier General William T. Wofford's infantry brigade and Brigadier General Williams C. Wickham's cavalry brigade across the Shenandoah to Guard Hill to defend the fords.

On the afternoon of August 16, and north of Guard Hill on the Winchester Road, Wickham's cavalry charged Brigadier General Thomas C. Devin's Federal vedettes at Cedarville. Devin counterattacked with two cavalry regiments and the action developed into a furious slashing fight with sabers. Wickham fell back, regrouped, and led another charge to allow his regiments to cross the North Fork and fall back to the south side of the river.

Wofford's grayclads, however, remained on Guard Hill on the north side of the river and came under attack from Brigadier General George A. Custer's cavalry brigade. Armed with Spencer repeating rifles, Custer's Michigan regiments drove Wofford's men off the hill and into the river, where many drowned.

With confirmation of Anderson's presence on the Front Royal–Winchester road, Sheridan withdrew down the valley at nightfall and eventually fell back to Harpers Ferry. The retrograde movement convinced Early that Sheridan was as timid as every other Union commander sent to the valley. The general's erroneous assumption later proved costly.

OPEQUON BRIDGE AND SMITHFIELD: AUGUST 29, 1864

As soon as Sheridan withdrew from Fishers Hill and returned to Harpers Ferry, Early began feeling his way down the Valley Pike, skirmishing with Sheridan's rear guard. After reaching Winchester on August 18, Anderson pushed on to Opequon Creek, which flowed north and passed to the east of Winchester, Smithfield, and Bunker Hill until emptying into the Potomac River. By August 19, Early had two of his divisions at Bunker Hill, located about twenty miles west of Harpers Ferry. On August 20 Early pressed eastward toward Charleston, ten miles closer to Harpers Ferry, without seeming to realize he had entered an area of severe danger because of his misconception that Sheridan would not fight. Early was unaware that on August 13 Grant believed that Lee had detached Longstreet's corps to support the Army of the Valley and had ordered Sheridan to fight defensively. Nor did Early realize that Sheridan was merely luring the Army of the Valley north to spring a giant trap.

For four days, Early remained in the vicinity of Charlestown, countering cavalry probes from the enemy while threatening Harpers Ferry. On August 25 he extended his line by going to Shepherdstown, and then to Martinsburg and Williamsport. Feeling increasing pressure from Federal cavalry and sensing a trap, Early returned to Bunker Hill. Federal cavalry began pressing across Opequon Creek near Smithfield but always fell back without bringing on an engagement. Early remained puzzled, but on August 29 the enemy came in greater force and drove the Confederates across the Opequon. Ramseur pressed forward while Gordon came around by back roads and struck the enemy's flank. Early seemed satisfied that the enemy would not fight but grew nervous and over the next several days moved back to the Winchester area. The wisdom of the retrograde movement became manifest two weeks later.

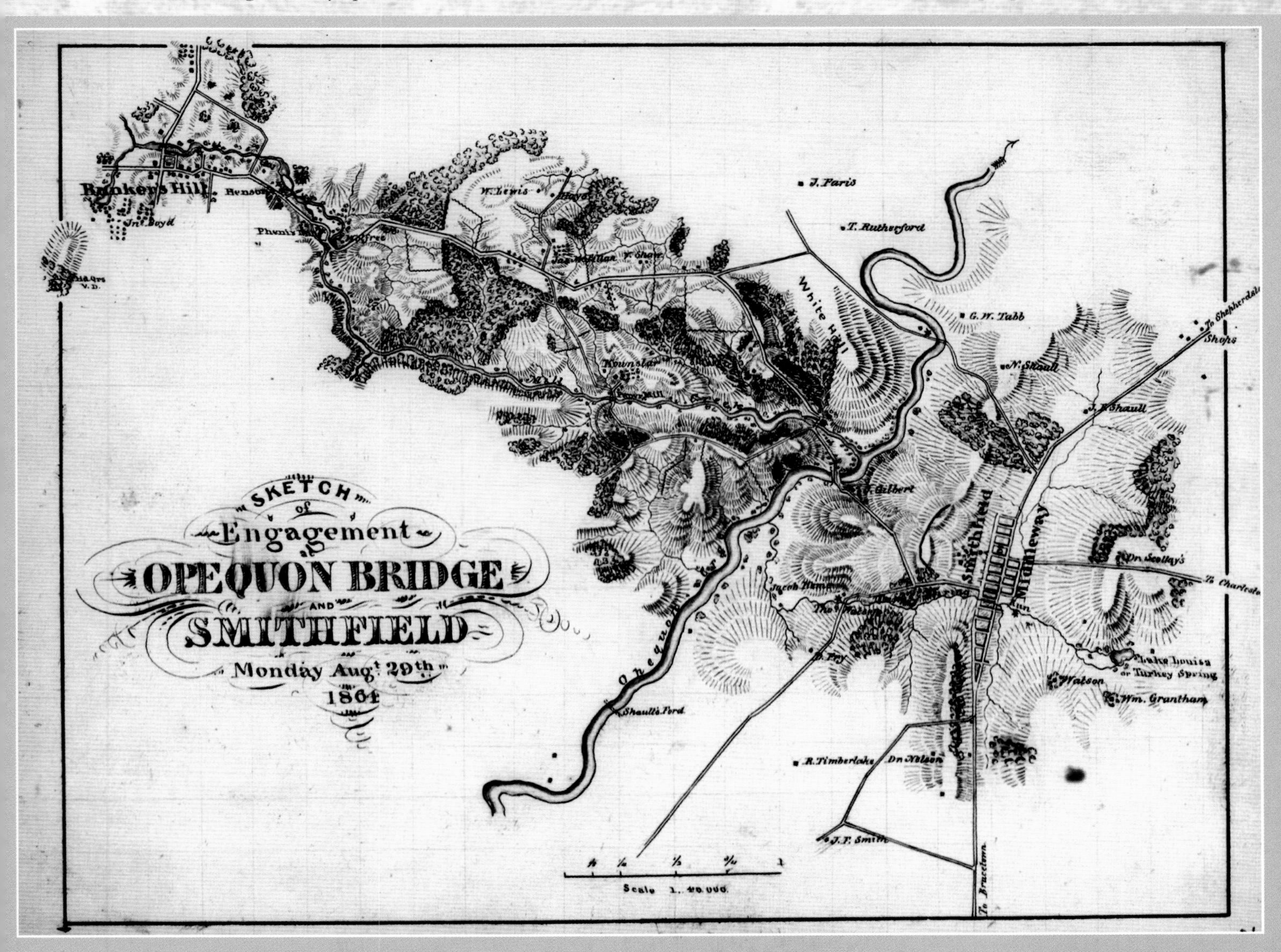

Engagement at Opequon Bridge [Sketchbook Map #8-22]
The engagement at the Opequon Bridge near Smithfield was typical of Sheridan's strategy to keep Early's army nearby and contained. Hotchkiss probably drew the map because he had time, and the engagement served as the only threat to the Army of the Valley since Fishers Hill on August 16.

THE THIRD BATTLE OF WINCHESTER: SEPTEMBER 19, 1864

In early September, Sheridan aligned his forces between Charlestown and Berryville, the latter lying ten miles east of Winchester, with Opequon Creek in between. During the same period, Kershaw's Confederates began to move toward Berryville and encountered infantry from Crook's Eighth Corps. After they surprised each other, a brisk fight ensued, thus forcing Kershaw back to Winchester. The next morning, Early came on the field planning to assault Crook's corps. When he saw Sheridan's entire force running north from Berryville, he pulled back and began to consolidate his brigades at Winchester. For the next ten days the two armies, six miles apart with Opequon Creek in between, each waited for the other to advance.

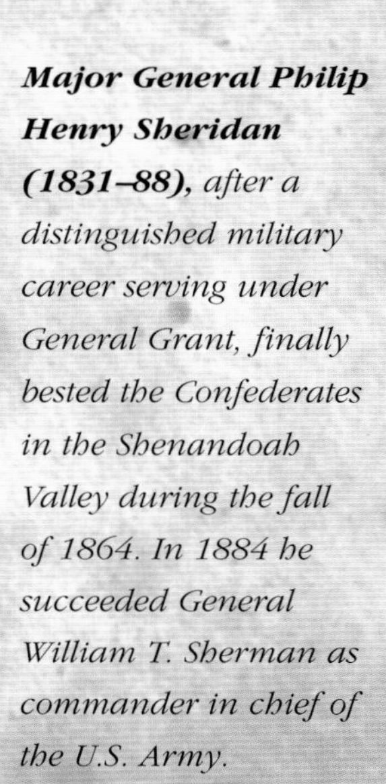

Major General Philip Henry Sheridan (1831–88), *after a distinguished military career serving under General Grant, finally bested the Confederates in the Shenandoah Valley during the fall of 1864. In 1884 he succeeded General William T. Sherman as commander in chief of the U.S. Army.*

On September 16, during the standoff at Opequon Creek, General Grant arrived at Harpers Ferry to meet with Sheridan. He carried a plan in his pocket, which he intended to give to Sheridan. Instead, Grant listened as Sheridan explained his own strategy. Becoming more impressed with Sheridan's plan than his own, Grant never pulled the document from his pocket and merely said, "Go in."

Once again, Early misread Sheridan's reluctance to fight as timidity and on September 17 began dividing his army for another raid down the valley, leaving only two divisions at Winchester. At nightfall on September 18, Gordon's division camped at Bunker Hill, thirteen miles to the north; Breckinridge went into camp with Rodes five miles away near Stephenson's Depot; Brigadier Gabriel C. Wharton's undersize division camped between Stephenson's Depot and Winchester; and Ramseur's division straddled a creek two miles east of Winchester. Fitzhugh Lee's and Lunsford L. Lomax's cavalry divisions each covered a flank. As both armies settled for the night, Early's force consisted of about 13,000 infantry and 2,900 cavalry as opposed to Sheridan's 33,600 infantry and 6,400 cavalry.

Sheridan used his cavalry to initiate a concentric assault on Early's forces at Winchester. The cavalry moved at dawn on September 19, with Merritt's and Averell's divisions crossing Opequon Creek at daybreak and attacking Early's flank at Stephenson's Depot north of Winchester. Brigadier General James H. Wilson's Federal cavalry division crossed Opequon Creek south of Winchester and pushed Lomax's and Jackson's cavalry back toward the town.

Major General John Cabell Breckinridge (1821–75) *moved his small command into the Shenandoah Valley in June 1864 and combined his forces with those of General Early. After the Washington raid and the battle of Winchester, Breckinridge became the Confederate's secretary of war.*

Early quickly deduced Sheridan's plan and recalled Rodes, Breckinridge, and Gordon from Stephenson's Depot for the defense of Winchester.

As Federal cavalry nibbled at the flanks of the Confederates defending Winchester, Wright's Sixth Corps struck Ramseur's division at 5:00 a.m., followed by Emory's Nineteenth Corps and Crook's Eighth Corps. Ramseur fell back from the pressure. Gordon arrived in time to slow Crook's assault, and Rodes to rebuff Emory's assault. Breckinridge arrived in midafternoon, barely in time to check Crook's flank attack. At 11:00 a.m., Rodes and Gordon observed a gap between the Federal lines and counterattacked. Rodes was killed, which took some of the steam out of the assault. The battle swayed back and forth throughout the afternoon. By 5:00 p.m., Early's infantry had been compressed into a defensive perimeter on the northeastern outskirts of Winchester. Fearing complete envelopment, Early ordered a general withdrawal to Fishers Hill at nightfall.

Expressing no particular bravado, Sheridan sent Grant a telegram announcing the victory at Winchester. Grant knew what it meant. "He came out of his tent," an observer recalled, "threw his hat in the air, and went back in again. He knew that was the beginning of the end."

Major General Stephen Dodson Ramseur (1837–64), *after suffering three serious wounds earlier in the war, was mortally wounded at Cedar Creek on October 19, 1864. Hotchkiss and Ramseur spent many hours together reconnoitering and plotting strategy with General Early.*

THIRD BATTLE OF WINCHESTER CASUALTIES

	Engaged	Killed	Wounded	Missing	Total
Confederate	16,377	276	1,827	1,818	3,921
Federal	37,711	697	3,983	338	5,018

THE BATTLE OF FISHERS HILL: SEPTEMBER 22, 1864

Although he copied a Yankee map a month later, Hotchkiss never made a map of the battle of Winchester because Early had given him permission to go home for a few days. Hotchkiss did not learn the extent of the Confederate defeat or of Rodes's death until he returned on September 20 and met long lines of ambulances and wounded grayclads limping through the gap in Fishers Hill on the Valley Turnpike. "We spent the day in line of battle and fixing our works," Hotchkiss wrote the following day. "The enemy's infantry in front and cavalry on the left made some demonstrations." Hotchkiss assessed the situation correctly by assuming there would be heavy fighting in the morning.

Early placed 9,000 effectives on Fishers Hill, not counting the cavalry distributed on the flanks. The crest twisted and turned for more than two miles, and Early lacked the men and reserves to defend the position adequately. He gave Lomax's undermanned and demoralized dismounted cavalry at Little North Mountain perhaps the most difficult task of all. The infantry divisions on the ridge welcomed a fight, but the problem of firing too high when in an elevated position promised to become an issue again.

Having whipped Early at Winchester, the Federals arrived at Strasburg late on September 20 with reinvigorated morale. Sheridan had been stopped at Fishers Hill on August 13 and knew the problems associated with a frontal attack. During a late-night conference between Sheridan and his corps commanders, Crook suggested a flank attack on the Confederate left. Although Wright and Emory disagreed with Crook's scheme, Sheridan approved it. He capitalized on Crook's plan by ordering Major General Alfred T. A. Torbert to take two divisions of cavalry, sweep around Massanutten Mountain to New Market Gap, and block Early's retreat.

On September 21, in preparation for the following day's assault, Wright and Emory began moving their divisions into position for a frontal attack. Early in the morning of September 22, Crook snaked two divisions on a concealed route and entered the woods on the far side of Little North Mountain. The long, circuitous route, combined with scaling the mountain, took most of the day. Crook did not get his men into position until late afternoon.

Lomax had been engrossed throughout the day by the skirmishing and artillery exchange in front of Fishers Hill. When Crook's blueclads charged out of the woods on Lomax's left and rear, the dismounted Confederate cavalry ran and dissolved up the hill. Ramseur tried to shift his front but had too little room to maneuver. The attack unnerved the Confederates, and when Wright's and Emory's divisions began to charge the hill, Early's line collapsed. Nightfall saved the Confederates from complete disaster. Early evaded Torbert and took the remnants of his army into the Blue Ridge, but Sheridan regained control of the Shenandoah and crippled the Army of the Valley in two battles over a span of four days.

Fishers Hill cost the Federals only 500 casualties, of which only fifty-two were killed. Early lost more than 1,500 men, of which 1,100 were taken prisoner. Hotchkiss lost one of his closest friends, Sandie Pendleton, who suffered a mortal wound while they were both rallying the rear guard during the retreat. "The rout of wagons, caissons, limbers, artillery, and flying men was fearful as the stream swept [up] the pike toward Woodstock," wrote Hotchkiss. "The troops marched all night."

Battle of Fishers Hill
[Sketchbook Map #8-26]
Fishers Hill overlooked Strasburg and provided a natural stronghold with a high, steep bluff that was ideal for emplacing artillery and which had excellent cover for infantry. Flanked by Little North Mountain on the west and Massanutten Mountain on the east, Fishers Hill dominated the Shenandoah Valley at its narrowest point. With adequate manning, the position could be made impregnable. The western end of the ridge gradually sloped into a low valley and provided the least defensible section of the hill. Early was aware of the problem and strung Lomax's dismounted cavalry on the left flank and placed Ramseur's division nearby to defend the crest. Although the ridge could be strongly defended from the front facing north, the left flank and rear remained vulnerable.

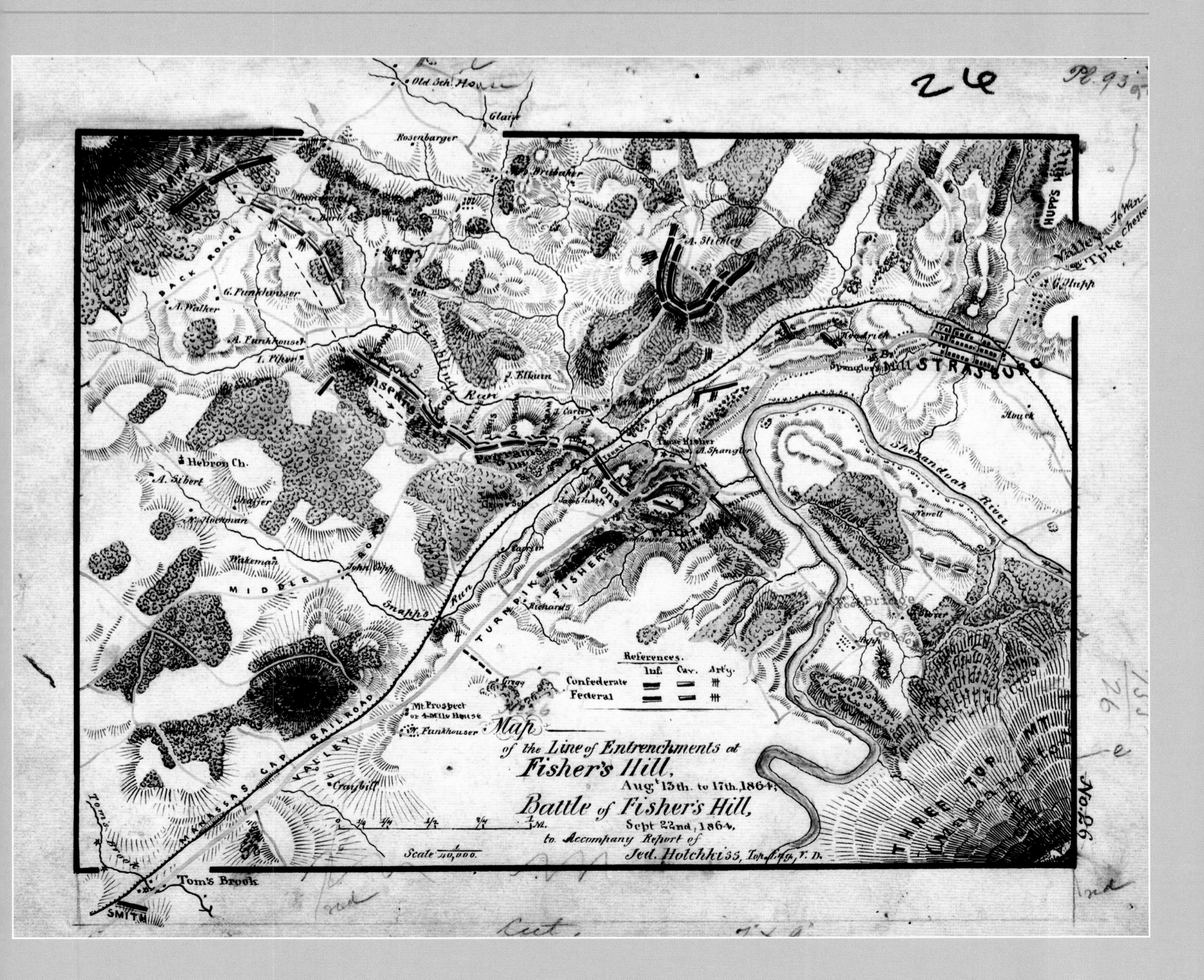
Map
of the Line of Entrenchments at
Fisher's Hill,
Aug't 13th. to 17th. 1864;
Battle of Fisher's Hill,
Sept 22nd, 1864,
to Accompany Report of
Jed. Hotchkiss, Top. Eng., V. D.
References.
Inf. Cav. Art'y.
Confederate
Federal
Scale 1/40,000
STRASBURG
Shenandoah River
Tom's Brook
MANASSAS GAP RAILROAD
VALLEY TURNPIKE
MIDDLE ROAD
BACK ROAD
Hebron Ch.
Tumbling Run
Snapp's Run
HUPPS HILL
THREE TOP MT.

Major General John Brown Gordon (1832–1904) *became one of Hotchkiss's most respected fighting generals. Gordon had been wounded five times, and when struck by a ball during the Shenandoah campaign, Hotchkiss remarked, "He never got off his horse to have his wound dressed, but pressed on after the enemy, all sprinkled with blood, the very personification of a hero."*

Sheridan relentlessly pursued the Army of the Valley to Rude's Hill, and then to Harrisonburg and Port Republic, setting fire to barns and crops as the blueclads advanced. By September 28, Early's shattered army reached Waynesboro, where the pressure finally subsided.

THE SMOLDERING VALLEY

Sheridan and Grant believed that Early's demoralized army had been whipped and no longer represented a threat, but both generals miscalculated. On October 5 Sheridan received approval from Grant to withdraw to Strasburg. The following day, blueclads formed two columns twenty miles long and marched north using roads on both sides of Massanutten Mountain. Sheridan lost track of Early's army and believed the remnants had scattered across the Blue Ridge. Instead, the Confederates had reorganized around Waynesboro and cautiously reconnoitered the Federal withdrawal.

What the Federal cavalry had not torched beforehand, they set afire during the course of their departure. Averell's division worked down the Luray Valley with his firebrands. Merritt's division burned farm buildings and granaries on the east side of the Valley Turnpike, and Custer's division scorched the west side of the pike. Sheridan later reported burning 2,000 barns and 500,000 bushels of grain while confiscating 50,000 head of livestock. For generations to come, the area's inhabitants would remember the events of the fall of 1864 as "the Burning."

"The enemy left Harrisonburg last night," wrote Hotchkiss on October 6. "We followed early this morning [with] our cavalry." By 11:00 a.m., Early had his army back on the road with Gordon's division in the van. Early did not say where he intended to go, but the indomitable general raged at Sheridan's wanton destruction of the valley. Clouds of smoke could be seen emitting from hundreds of barns and granaries, and at night an orange hue from burning buildings and haystacks lit up low, overhanging clouds for miles. Everyone knew from the voluble swearing of their feisty general that he wanted revenge. Early's smaller army had one advantage: Sheridan believed the Confederate force, except for a few regiments of cavalry and infantry, had either been destroyed or was no longer in the valley.

Battle of Cedar Creek [Sketchbook Map #8-29] *Cedar Creek (also called Belle Grove) represented the last major battle fought by the Army of the Valley. Hotchkiss helped plan the battle and was all over the field when the fighting began. He had scouted the enemy's position and mapped the routes for the attack. After the war, he revised his Cedar Creek map to include the positions and movements of every Confederate brigade in the battle, and that map appears only in* The Official Military Atlas of the Civil War.

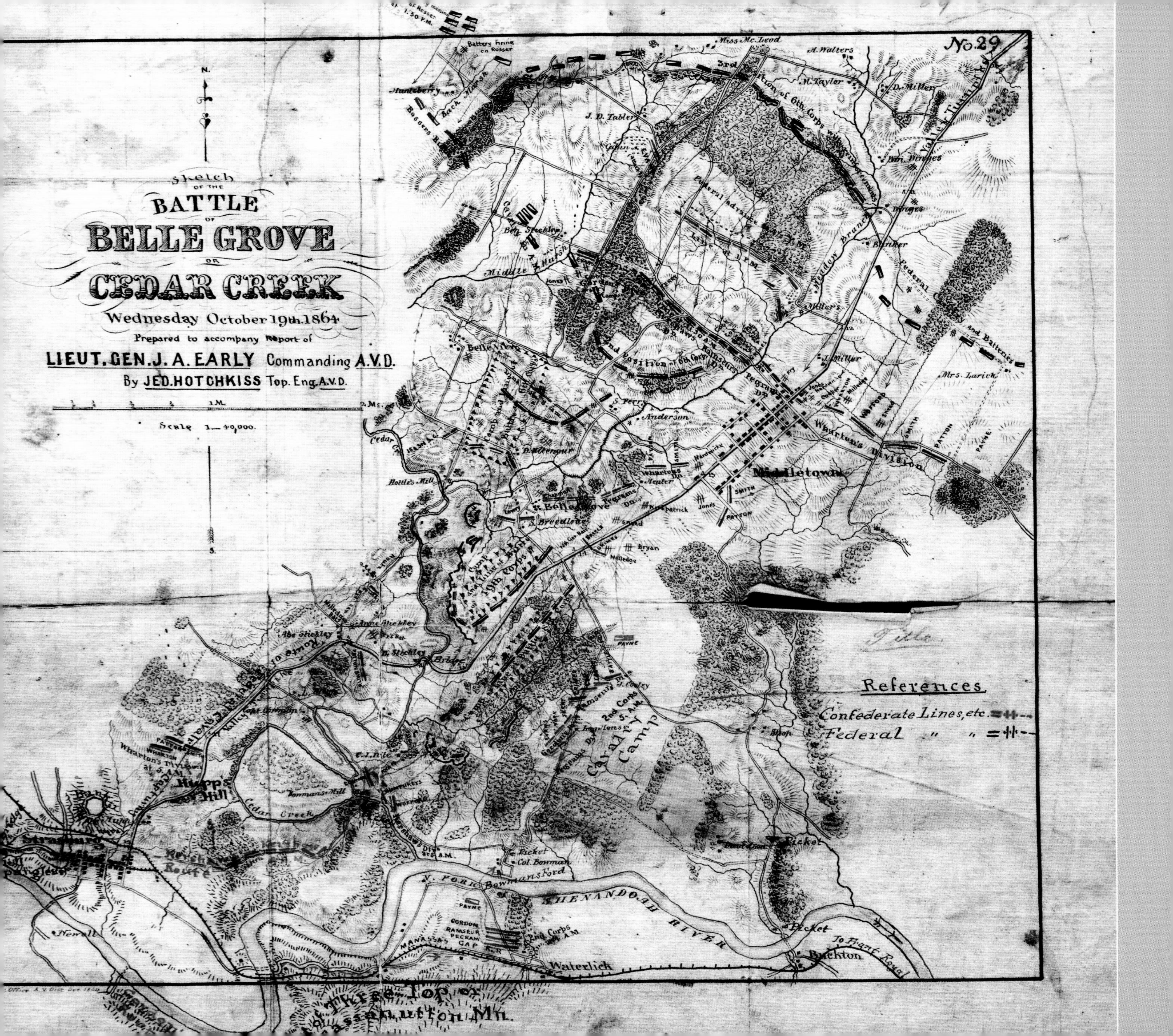
No. 29
Sketch of the BATTLE of BELLE GROVE or CEDAR CREEK
Wednesday October 19th. 1864
Prepared to accompany Report of
LIEUT. GEN. J. A. EARLY Commanding A.V.D.
By JED. HOTCHKISS Top. Eng. A.V.D.
Scale 1—40,000
N.
S.
References
Confederate Lines, etc.
Federal " "
Middletown
Belle Grove
Cedar Creek
Hupp's Hill
Strasburg
Bowmans Mill
Bowmans Ford
N. Fork
SHENANDOAH RIVER
Waterlick
Buckton
To Front Royal
Wharton's Division
Middle Road
Valley Turnpike
Meadow Brook
Cavalry Camp
2nd Corps
Picket
Col. Bowman
Manassas Gap R.R.
Belleview
Hottle's Mill
J. D. Tabler
Miss Mc. Leod
A. Walters
M. Taylor
D. Miller
J. Miller
Mrs. Larich
Anderson
Bryan
Milledge
Newell
Three Top or Massanutton Mn.

THE BATTLE OF CEDAR CREEK: OCTOBER 19, 1864

Screened by cavalry, Early's Army of the Valley moved north, using back roads and gaps in the mountains to remain concealed. On October 13 Lomax's brigade reported the Federals encamped around Belle Grove at Cedar Creek, which was located between Strasburg and Middleton on the Valley Turnpike. Using cavalry to distract the enemy, Early quietly put his infantry back on Fishers Hill. On October 18 Hotchkiss participated with all the division commanders, including artillery and cavalry commanders, in a strategy session designed to drive the Federals out of the valley. Several routes of attack came under discussion, but Early decided to use a brilliant strategy proposed by Hotchkiss and General Gordon that would bring the entire Confederate force against the flank of the Federal position. The plan having been decided, Hotchkiss, Gordon, and Ramseur climbed to the signal station on Three Top Mountain. Using a strong glass, they studied the entire Federal camp and the surrounding terrain. Once again they discussed the route.

After returning to camp, Hotchkiss led a detachment of pioneers along the designated route to remove trees and build a footbridge across the North Fork of the Shenandoah at Fishers Hill, thereby enabling the infantry to cross swiftly. Gordon's divisions took the lead at 8:00 p.m. and marched through the night to get into position for the morning attack.

Sheridan had deposited about 31,000 Federals in the area of Cedar Creek, which included elements from Wright's Sixth Corps, Emory's Nineteenth Corps, and Crook's Eighth Corps. Sheridan was totally unaware that Lee had reinforced Early, whose Army of the Valley had sprouted to about 21,000 effectives. A great many of the wounded had returned, filling slots in Early's five infantry divisions and three cavalry brigades.

Hotchkiss recalled that near dawn a light mist hung over Cedar Creek and the river. He had recommended some adjustments in the position of the infantry, all of which were made during the night. The three divisions led by Gordon, Ramseur, and Pegram filed into position to strike Crook's rear at dawn. Kershaw's division waited on a side road while Wharton's division waited beside the Valley Turnpike. Early also detailed a small cavalry force to ride into the Federal camp and capture Sheridan, who was not there but in Winchester, having that night returned from a meeting in Washington.

Before daybreak on October 19, Early's five infantry divisions charged through the morning mist and struck Crook's sleeping camp. The assault spread to Wright's camp, where the rout of the Sixth Corps rolled through the camp of Emory's Nineteenth Corps. Wright failed to make a stand and the Sixth Corps became commingled with Emory's corps in the rush to escape from the rebel-yelling Confederates. By 10:00 a.m., the Federals stopped in disorder in the farmlands beyond Middletown and began to regroup. The Confederate assault ran out of steam, and Early formed a new line outside Middletown.

For the next six hours, Confederates in the rear looted Federal camps, consuming rations and gathering tons of supplies. A thousand or more grayclads, fatigued from an all-night march and desperately hungry, broke formation and stopped to enjoy a meal at the enemy's expense. Early tolerated the looting because he needed supplies, but he also experienced difficulty reconsolidating his forces, which had run wildly through the enemy's camps. His splendid victory seemed almost assured, but then—unfortunately—he decided to pause. During the morning, Gordon urged Early to exploit his advantage and press the attack before the much-superior enemy reorganized and counterattacked, but the Confederate force had become a jumbled mass of disassociated units.

Sheridan awoke early at Winchester and heard the faint booming of artillery coming from the Middleton area. He galloped up the pike toward the firing, and soon encountered long lines of Federals streaming back toward Winchester. At 10:30 a.m., he arrived on the field and found Wright, who had been left in charge, dazed by the assault. Sheridan rode through his bewildered army, inspiring the men to re-form. At 4:00 p.m., he ordered a counterattack, regained the initiative, and drove the Confederates back to Fishers Hill with heavy losses. Sheridan lost 5,665 men, including about 1,600 missing. Early lost about 2,900 men, most of his artillery, all his ammunition wagons and ambulances, and General Ramseur, who died in battle.

Hotchkiss summed up the disaster, writing, "Thus was one of the most brilliant victories of the war turned into one of the most disgraceful defeats, and all due to the delay in pressing the enemy after we got to Middletown." That evening, Hotchkiss remembered Early as being "very much prostrated when he learned the extent of our disaster." Hotchkiss added, "The Yankees got whipped, and we got scared." Early never learned of Sheridan's involvement in the disastrous reversal until later.

LIFE IN CAMP

On October 20 the Confederates retired to New Market. The men of the Army of the Valley were not ready to admit that for them it was over. A few days later, Hotchkiss traveled to Richmond to deliver Early's report to Lee. In his journal, Hotchkiss wrote, "General Early told me not to tell General Lee that we ought to have advanced in the morning at

Middletown, for, he said, we ought to have done so." A few days later, Hotchkiss returned to New Market and spent most of the following months making maps, including a more comprehensive map on the battle of Cedar Creek. Early's small army continued to have skirmishes and mini battles with Sheridan's cavalry, but the Confederates were contained and could do little more than defend themselves until it ended.

WINTERING IN THE VALLEY

Hotchkiss spent much of the fall and winter near Staunton, close to his home. Although Early occasionally moved his rapidly shrinking army up and down the valley, most of the fighting began and ended in cavalry skirmishes. Hotchkiss was well aware that Early's losses in the valley, coupled with the fall of Atlanta, ensured the reelection of Lincoln and the continuance of the war. The interregnum in the fighting gave Hotchkiss, Robinson, and Oltmanns time to finish dozens of maps that had been started and never completed, and most of the maps pertaining to 1864 operations from the Wilderness to Cedar Creek fell into that category.

One day blended with the next. The most unusual news came in mid-November when Hotchkiss learned he had been promoted to first military assistant engineer with an annual salary of 4,000 worthless Confederate dollars. This was the highest civilian position in the Engineer Corps and also gave Hotchkiss the equivalent rank of major, though he never officially became an officer in the Confederate army.

For the next four months, Hotchkiss recorded the pattern of his work in his journal. November 17: "Worked at battle of Cedar Creek; Robinson finished Southeast Virginia." November 28: "Finished battle of Cedar Creek . . . and worked on battle of Kearneysville. Oltmanns finished Western Virginia; Robinson made map of Harper's Ferry." Hotchkiss even received calls from other fields of battle, writing on December 2, "Oltmanns finished map of battle of Chattanooga." New Year's Day came and went, and on January 17 Hotchkiss wrote, "Finished Wilderness map and began one of Spotsylvania battles." For Hotchkiss, every map remained in transition until he had no more to add.

Work abruptly halted on March 1, 1865, on reports of Federals advancing toward Staunton. The operation would be Sheridan's last advance up the valley, bringing 10,000 blueclads in two divisions from Merritt's cavalry and one infantry division. Early pulled back to Waynesboro with fewer than 2,000 men, all others having been transferred to Lee's army at Petersburg. Early tried to rally his forces, but they were too few. In the rout that followed, Early lost the last remnants of his army. Sixteen hundred Confederates surrendered, relinquishing fourteen cannons and 200 wagons. Hotchkiss wrote on March 3, "The whole army scattered." When he rode into Waynesboro, the town was filled with Federals, so he spurred his horse onto a winding trail up the side of a mountain.

After the blueclads passed through Waynesboro and started toward Charlottesville, Hotchkiss descended from his mountain refuge and went home, wondering how much longer the war could last. Learning that Early had left the valley and gone to Richmond, Hotchkiss started toward the capital on March 11 with General Rosser and about 500 men. He reached Richmond on March 16, only to be told to return to Staunton. After a tortuous journey on a rickety wagon drawn by four mules, he reached Staunton on March 28 and went home. With the state of affairs unknown, Hotchkiss later returned to his office and resumed making maps with Robinson.

Learning on April 4 that Richmond had been evacuated, Hotchkiss went with General Lomax's cavalry to Lynchburg. A few days later, Hotchkiss heard that Lee had surrendered on April 9 at Appomattox Court House. He never reached Appomattox. He simply turned around and started back to the valley. By then, the roads were filled with parolees trudging home to begin the spring planting. A few days later, he learned that Lomax and Jackson had disbanded their brigades, so on April 16 he disbanded himself and brought Robinson back to Loch Willow to stay with the family. When a Federal provost marshal appeared in Staunton on May 1, Hotchkiss stood in line with a group of Confederate soldiers and collected his official parole.

The Civil War ended in defeat for the South, but for Hotchkiss, now freed from the shackles of war, there would be a new beginning. His talents would not go unused in a new, united nation springing forth from the miseries of war.

EPILOGUE

The war had been an unexpected education for Jedediah Hotchkiss. He suffered through it, every day homesick for his wife and two daughters. As Hotchkiss was a devoted family man, his decision to go so quickly to war was curious. He left behind a promising life as an educator and a landowner of substance.

On April 18, 1865, Hotchkiss surprised his family when he unexpectedly arrived at Loch Willow at dusk. The Soldier's Aid Society of Churchville had just adjourned a meeting at his home, and Hotchkiss felt indebted to them. The good people of Churchville had provided for his family and for hundreds of others whose fathers and sons had served the Confederacy.

Major Jedediah Hotchkiss (1828–99) *used the Confederate army as a training experience to perfect skills that carried him through life during the postwar years. Working with trained engineers like James Boswell added to his knowledge of engineering, which led to a productive career in the years after the war.*

Though debts had stripped away most of Hotchkiss's possessions, they had not broken his spirit. He had physically suffered while loyally serving the South, and now it was over. Still a young man at thirty-six, he did not look back. His spirit had not been broken by war, nor had his faith. Capitalizing on his experiences, he buried his remorse and looked straight ahead.

War had taken a toll on his family. They were hungry, weary from suffering, and worried they might be ejected from their rented dwelling at Loch Willow. Hotchckiss settled accounts with a small amount of money, dug a garden, and planted enough vegetables to see the family through the summer and early fall. There were no jobs, but the valley had to be rebuilt; there were few schools, but there were children who needed to be educated. Hotchkiss wasted no time formulating plans to organize classes. He fixed up his schoolroom, which had been neglected during the war, and on May 10 enrolled the first four students in Loch Willow Academy. Each paid $100 tuition for a ten-month term. On June 10 he closed the school for the summer, and with financial burdens piling up, he looked for work until classes resumed on September 1.

NEW OPPORTUNITIES

Hotchkiss received a letter in October from William Allan, a comrade who had served as Stonewall Jackson's ordnance chief, stating that Colonel Michael G. Harman, who had commanded the Fifty-second Virginia in the Valley Army, wanted his acreage surveyed. Allan implied that this would lead to other work. Michael Harman was the brother of John Harman, who had served as Jackson's chief quartermaster. In addition to having his track surveyed, Harman also wanted Hotchkiss to move to Staunton and open an exclusive school for fifteen boys. Hotchkiss would receive $100 a year per student plus an additional $540 for Harman's two children. Hotchkiss sealed the deal, moved his family to Staunton, and on September 6 opened the school in a long building behind his home. When not teaching, Hotchkiss filled his time as a surveyor and on occasion lectured at other schools.

In October a Federal agent from Richmond arrived in Staunton and demanded Hotchkiss's wartime map collection. Hotchkiss refused to surrender the maps, claiming they did not fall under the category of "Confederate documents" and were private property, but he promised to carry them to Richmond to protest the seizure. After Federal agents in Richmond refused to bargain, Hotchkiss packed his maps and went to Washington. He convinced Grant of the historical value of the collection, and the general agreed to leave the maps in the possession of Hotchkiss in exchange for the latter's agreement to provide copies. What began as a potential devastating blow to Hotchkiss and his cherished collection worked to his advantage. Grant selected a number of maps he wanted copied and agreed to pay for them. Hotchkiss hired Oltmanns to produce the copies Grant requested. After meeting with Grant, the government continued to request the Virginian's help in compiling the records of the 1864 Shenandoah Valley campaign for *The War of the Rebellion: A Compilation of the Official Records of the Union and Confederate Armies,* in which portions of Hotchkiss's journal appears.

Colonel William Allan (1837–89) *served as General Jackson's ordnance chief and once warned the loquacious Hotchkiss, "Talking will be the death of you yet." The two men, living in the same area of the valley, became close friends and fellow staff members. After the war, Allan became instrumental in several phases of Hotchkiss's life, which led to the latter becoming a civil and mining engineer and an author.*

In 1866 Hotchkiss became involved in projects that began to take him out of him the classroom. Anyone in the area who needed an engineer came to Hotchkiss, and everybody from the old Valley Army who wanted to write reminiscences also came to Hotchkiss. When William Allan began to write *The Battlefields of Virginia: Chancellorsville*, he offered Hotchkiss two-thirds of the profits in exchange for the use of the latter's maps and journal. Published in 1867 and now worth about $2,000, the book came out in Hotchkiss's name with Allan as coauthor. Allan later wrote two more books with Hotchkiss's help. *The Battlefields of Virginia: Chancellorsville*, which actually included several other battles, represented one of the earliest books on the Civil War written by two intelligent Southerners who had been with the Stonewall Brigade since its inception.

ENGINEER AND PROMOTER

By 1867 Hotchkiss abandoned the classroom and devoted his time to engineering and lecturing. In the months that followed, he opened an office in Staunton, advertising himself as "Jed Hotchkiss: Mining and Consulting Engineer." No other Virginian had a more thorough knowledge of the commonwealth's topography, geology, and untapped mineral resources than Hotchkiss. He used this knowledge to attract capital from the North to exploit Virginia's coal and mineral resources and began writing books, articles, and reports on his research. In 1867 he personally launched a campaign for the commonwealth's restoration and, with his brother

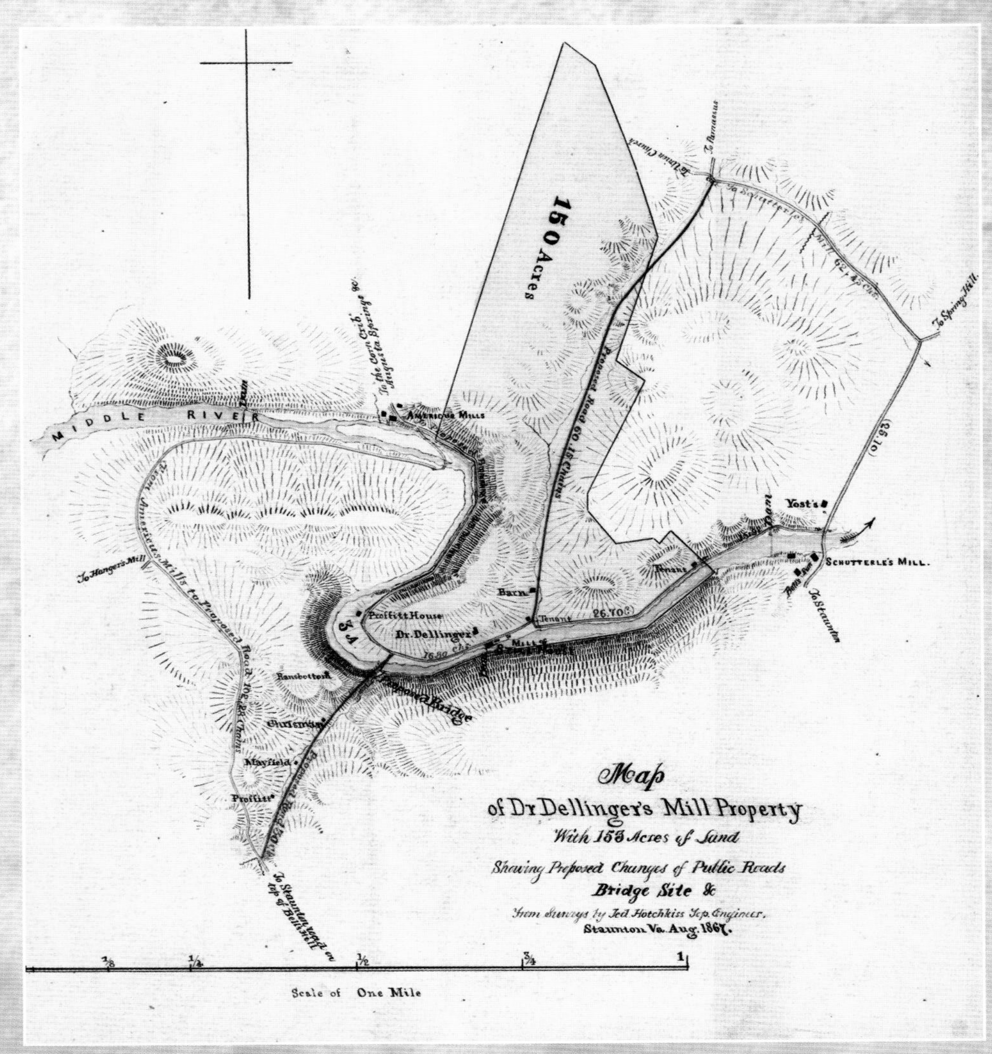

Dr. Dellinger's Mill Property [Map #312] *One of Hotchkiss's early engineering assignments involved the property of Dr. Dellinger, who owned a large field of sand used in the foundry and glass industries. Hotchkiss surveyed the land and laid out two roads and a bridge for getting the sand transported to the Virginia Central Railroad in Staunton.*

Nelson, spent the next thirty years promoting Virginia's attributes. When he could not raise money in Virginia or in the North, he raised it in England, even though the British still looked upon the United States as a politically unstable nation.

Hotchkiss took particular interest in developing the resources of Rockbridge, Augusta, and other western Virginia counties not far from his home in Staunton. He developed mining interests and produced proposals that would provide work for thousands of able-bodied African Americans who were anxious to improve their lives. With Nelson, who was already connected with the Chesapeake and Ohio and the Richmond and New York River railroads, Hotchkiss brought new rail service into the area to transport the valley's grain and mineral resources to the populous cities of the East.

RECORDING THE WAR

During the early 1880s, Hotchkiss noticed a huge interest developing among the public for histories about the Civil War. Literary activity provided a stimulus for Hotchkiss, who wrote *Virginia*, the third volume of General Clement A. Evans's *Confederate Military History*. Evans had served under Jackson, Early, and Gordon, and prior to the battle of Cedar Creek he had gone to the top of Massanutten Mountain with Hotchkiss and two other officers to reconnoiter the Federal position. The entire series, and Hotchkiss's contribution in particular, became best sellers.

Hotchkiss's map collection continued to be a source of interest. In 1885 he copublished with Joseph Waddell the *Historical Atlas of Augusta County*, which contained maps made during and after the Civil War. Although *The War of the Rebellion* was published in seventy volumes between 1880 and 1901, the maps had been removed from the battle reports and put in the *Atlas to Accompany the Official Records*, which was originally published in 1891–95. Today, reprints of the atlas are titled *The Official Military Atlas of the Civil War*. Hotchkiss redrew his sketches, converting them to the stylistic requirements for publication, and contributed about half of the Confederate maps and sketches that appear in the atlas.

Hotchkiss's involvement in Civil War literature led him to the lecture circuit across the Uited States; he gave three lectures titled "The Valley Campaign of 1862," "The Battle of the Wilderness," and "Topography in War." For each lecture, he carried a portable blackboard and colored chalk that was used for making extemporaneous sketches.

In the late 1880s and early 1890s, British military historian G. F. R. Henderson came to the United States to work on a two-volume biography on Stonewall Jackson. Henderson became connected with Hotchkiss after he learned the topographical engineer had written *Virginia* as part of *Confederate Military History*. Henderson wanted to know all about Jackson's character, personality, idiosyncrasies, command of tactics, and how dependent the general had been on maps. Only Hotchkiss could relate Jackson's dependence on maps and how the general performed with them as opposed to how he performed without them. Henderson soon concluded that Jackson's successes depended upon Hotchkiss and the general's lapses at Romney and Richmond occurred because of Hotchkiss's absence. Henderson brought this factor to light in 1898 when he first published *Stonewall Jackson and the American Civil War*. His maps came mainly from copies provided by Hotchkiss. Every chapter Henderson wrote went first to Hotchkiss for criticism and correction. Hotchkiss numbered every suggestion for the manuscript and on a separate sheet of paper explained the reason why certain details should be inserted or changed. Fifty years passed before anyone produced a biography on Stonewall Jackson that approached the quality, thoroughness, and accuracy of Henderson's two-volume work. What made Henderson's biography so distinctly superior were the annotations from Hotchkiss, who painstakingly wrote letters to his old comrades to gather additional information and confirm what Henderson had written. Hotchkiss probably put more effort into ensuring the absolute accuracy of the work than did Henderson. The long correspondence between the two men became part of the Hotchkiss legacy, and those letters have been a priceless resource for Civil War historians ever since.

Hotchkiss lived comfortably throughout his later life without ever becoming wealthy. Always brimming over with warmth and kindness, he made many friends. Extremely talented as a self-taught engineer, Hotchkiss was also markedly inquisitive, thoroughly loyal, deeply religious, and charitable to those in need. He aged gracefully and his features barely changed except for the silvery flowing beard that he kept neatly trimmed.

Hotchkiss made it a practice, whenever possible, to never miss a service at the Second Presbyterian Church of Staunton, where he served as Sunday school superintendent. Finding himself delayed while traveling one Saturday, he attempted to return home by booking passage on a freight car running into Staunton late at night. By the time the train pulled into the station, Hotchkiss had become ill from exposure and jarring. Taken to Richmond later for surgery, he appeared to be recovering and returned home, but he died of mastoiditison on January 17, 1899.

Hotchkiss's death did not signal an end to his work, as it lives on in the history of the Civil War. He gave his great gift as a topographer to history, and he gave his life to the reconstruction of Virginia.

BIBLIOGRAPHY

Agassiz, George R., ed. *Meade's Headquarters, 1863–65: Letters of Colonel Theodore Lyman from the Wilderness to Appomattox.* Boston: Houghton Mifflin, 1922.

Alexander, E. P. *Military Memoirs of a Confederate.* New York: Charles Scribner's Sons, 1912.

Allan, William. *History of the Campaign of Gen. T. J. Jackson in the Shenandoah Valley of Virginia.* Dayton: Morningside Bookshop, 1974.

Bean, William G. *Stonewall's Man: Sandie Pendleton.* Chapel Hill, NC: University of North Carolina Press, 1959.

Bigelow, John Jr. *The Campaign of Chancellorsville.* New Haven, CT: Yale University Press, 1910.

Boatner, Mark M. III. *The Civil War Dictionary.* New York: David McKay Company, 1959.

Catton, Bruce. *Grant Takes Command.* Boston: Little, Brown, 1969.

Chambers, Lenoir. *Stonewall Jackson.* 2 vols. New York: William A. Morrow, 1959.

Coddington, Edwin B. *The Gettysburg Campaign: A Study in Command.* New York: Charles Scribner's Sons, 1968.

Davis, Burke. *Jeb Stuart: The Last Cavalier.* New York: Rinehart & Company, 1957.

Early, Jubal Anderson. *War Memoirs: Autobiographical Sketch and Narrative of the War Between the States.* Bloomington, IN: Indiana University Press, 1960.

Faust, Patricia L., ed. *Historical Times Illustrated Encyclopedia of the Civil War.* New York: Harper & Row, 1986.

Fox, William F. *Regimental Losses in the American Civil War.* Albany, NY: Albany Publishing Co., 1898.

Frassanito, William A. *Grant and Lee: The Virginia Campaigns 1864–1865.* New York: Charles Scribner's Sons, 1983.

Freeman, Douglas Southall. *Lee's Lieutenants: A Study in Command.* 3 vols. New York: Charles Scribner's Sons, 1942–1944.

________. *R. E. Lee: A Biography.* 4 vols. 1935.

Gordon, John B. *Reminiscences of the Civil War.* New York: Charles Scribner's Sons, 1903.

Grant, U. S. *Personal Memoirs of U. S. Grant.* New York: AMS Press, 1972.

Henderson, G. F. R. *Stonewall Jackson and the American Civil War.* 2 vols. New York: Longmans, Green & Co., 1898.

Hotchkiss, Jed. *Virginia.* vol. 3 of *Confederate Military History,* Clement A. Evans, ed. Atlanta: Confederate Publishing Company, 1899.

Hotchkiss, Jedediah. Papers, letters, diary, journal, and map collection. Library of Congress, http://hdl.loc.gov/loc.mss/eadmss.ms006031.

Johnson, Robert Underwood, and Clarence Clough Buel, eds. *Battles and Leaders of the Civil War.* 4 vols. New York: Thomas Yoseloff, Inc., 1956.

Krick, Robert K. *Lee's Colonels.* Dayton, OH: Morningside House, 1991.

LeGear, Clara E., comp. "The Hotchkiss Collection of Confederate Maps," *Library of Congress Quarterly Journal of Current Acquisitions,* 6 (November 1948), 16–20.

________. *The Hotchkiss Map Collection: A List of Manuscript Maps.* Falls Church, VA: Sterling Press, 1977.

Livermore, Thomas L. *Numbers and Losses in the Civil War in America 1861–1865.* Boston: Houghton Mifflin Co., 1901.

Longstreet, James. *From Manassas to Appomattox: Memoirs of the Civil War in America.* Bloomington, IN: University of Indiana Press, 1960.

McClellan, H. B. *The Life and Campaigns of Major General J. E. B. Stuart.* Boston: Houghton Mifflin Co., 1885.

McDonald, Archie P., ed., *Make Me a Map of the Valley: The Civil War Journal of Stonewall Jackson's Topographer.* Dallas: Southern Methodist University Press, 1973.

Miller, William J. *Mapping for Stonewall: The Civil War Service of Jed Hotchkiss.* Washington, D.C.: Elliott & Clark Publishing, 1993.

Pond, George E. *The Shenandoah Valley in 1864.* New York: Charles Scribner's Sons, 1889.

Porter, Horace. *Campaigning with Grant.* New York: The Century Company, 1897.

Robertson, James I. Jr. *General A. P. Hill: The Story of a Confederate Warrior.* New York: Random House, 1987.

________. *The Stonewall Brigade.* Baton Rouge, LA: Louisiana State University Press, 1985.

Roper, Peter W. *Jedediah Hotchkiss: Rebel Mapmaker and Virginia Businessman.* Shippensburg, PA: White Mane Publishing Co., 1992.

Schaff, Morris. *The Battle of the Wilderness.* Boston: Houghton Mifflin Co., 1910.

Sorrel, G. Moxley. *Recollections of a Confederate Staff Officer.* Jackson, TN: McCowat-Mercer, 1958.

Stackpole, Edward J. *Sheridan in the Shenandoah: Jubal Early's Nemesis.* Harrisburg, PA: The Stackpole Company, 1961.

Tanner, Robert G. *Stonewall in the Valley.* New York: Doubleday, 1976.

Taylor, Richard. *Destruction and Reconstruction.* New York: Longmans, Green & Co., 1955.

Thomason, John W. Jr. *Jeb Stuart.* New York: Charles Scribner's Sons, 1930.

The War of the Rebellion: A Compilation of the Official Records of the Union and Confederate Armies. 70 vols. Harrisburg, PA: National Historical Society, 1971.

Wise, Jennings Cropper. *The Long Arm of Lee: The History of the Army of Northern Virginia.* New York: Oxford University Press, 1959.

INDEX